WEYERHAEUSER ENVIRONMENTAL BOOKS

William Cronon, Editor

WEYERHAEUSER ENVIRONMENTAL BOOKS explore human relationships with natural environments in all their variety and complexity. They seek to cast new light on the ways that natural systems affect human communities, the ways that people affect the environments of which they are a part, and the ways that different cultural conceptions of nature profoundly shape our sense of the world around us. A complete list of the books in the series appears at the end of this book.

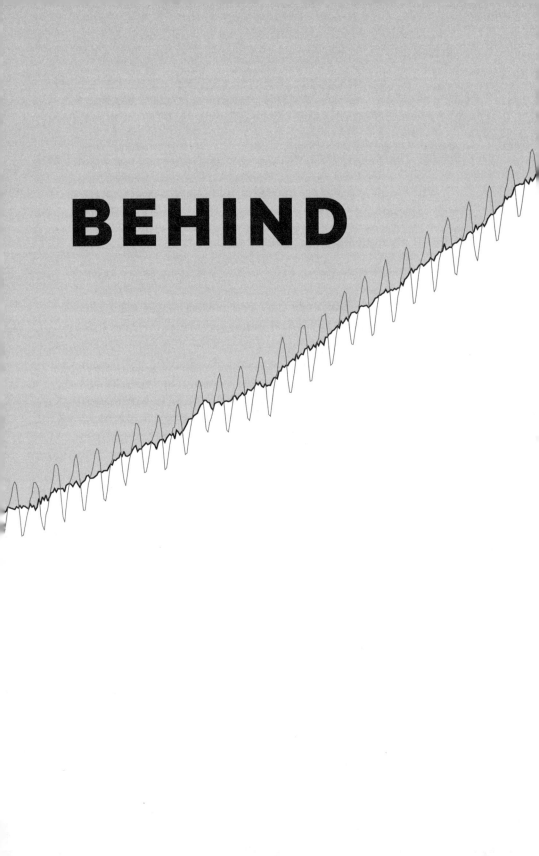

BEHIND

THE CURVE

SCIENCE AND THE POLITICS
OF GLOBAL WARMING

Joshua P. Howe

UNIVERSITY OF WASHINGTON PRESS
Seattle and London

34040876

Behind the Curve is published with the assistance of a grant from the Weyerhaeuser Environmental Books Endowment, established by the Weyerhaeuser Company Foundation, members of the Weyerhaeuser family, and Janet and Jack Creighton.

UNIVERSITY OF WASHINGTON PRESS
PO BOX 50096, Seattle, WA 98145, USA
www.washington.edu/uwpress

CATALOGING-IN-PUBLICATION DATA IS ON FILE WITH THE LIBRARY OF CONGRESS

ISBN 978-0-295-99368-3

The paper used in this publication is acid-free and meets the minimum requirements of American National Standard for Information Sciences—Permanence of Paper for Printed Library Materials, ANSI Z39.48–1984.♾

CONTENTS

FOREWORD

A Fateful Line

WILLIAM CRONON

IT HAS BECOME ONE OF THE MOST CONSEQUENTIAL STORIES OF THE twentieth century. In 1956, a young chemist named Charles David Keeling joined the staff of the Scripps Institution for Oceanography in La Jolla, California. Keeling had been developing new instruments for measuring minute concentrations of different gases in the atmosphere and had recently perfected a device capable of detecting carbon dioxide in quantities so small that they amounted to just a few hundred parts per million. The oceanographer Roger Revelle, who in his capacity as director of Scripps was responsible for recruiting Keeling, had been trying for some time to persuade his fellow scientists of the importance of monitoring gaseous interactions between the oceans and atmosphere. In the same year that he hired Keeling, Revelle authored with his colleague Hans Suess a paper with the dry academic title, "Carbon Dioxide Exchange Between Atmosphere and Ocean and the Question of an Increase of Atmospheric CO_2 during the Past Decades." Buried amid the thickets of tables, graphs, and formulae that make up most of the article are four sentences that have become one of the most frequently quoted passages from any scientific publication of that decade:

> Human beings are now carrying out a large scale geophysical experiment
> of a kind that could not have happened in the past nor be reproduced in
> the future. Within a few centuries we are returning to the atmosphere
> and oceans the concentrated organic carbon stored in sedimentary
> rocks over hundreds of millions of years. This experiment, if adequately
> documented, may yield a far-reaching insight into the processes deter-
> mining weather and climate. It therefore becomes of prime importance
> to attempt to determine the way in which carbon dioxide is partitioned
> between the atmosphere, the oceans, the biosphere and the lithosphere.

Just so did a story begin whose full implications few people at the time could have anticipated: Revelle had hired Keeling in the hope that the young man's new instruments might be just what was needed to "document" this "large scale geophysical experiment."

It was an auspicious moment to begin such an investigation. 1957-58 had been designated as the first-ever International Geophysical Year, during which scientists from all over the world, including the United States and the Soviet Union, would cooperate to investigate large-scale atmospheric, oceanographic, and geophysical processes that were receiving significant new attention (and funding) in the context of the Cold War. Suddenly, predicting the downwind movement of fallout and tracking minute quantities of radioactive isotopes in the atmosphere gained new urgency as military and civilian leaders sought to understand what might be involved in fighting a nuclear war. Although physics and chemistry typically take center stage in narratives about the making of the atomic bomb, it is less often noticed how important meteorology and climate science became in a world where that terrifying new weapon would be delivered by long-range bombers and ICBMs and where airborne particulates would be among the most lethal byproducts of the resulting explosions. Wind had never before seemed quite so apocalyptic a force in human history.

David Keeling began his work in 1957 by placing instruments at three locations: La Jolla, Antarctica, and Mauna Loa on the Big Island of Hawai'i. (Airborne sampling was initially also done in the Arctic by the 55th Weather Reconnaissance Squadron of the United States Air Force, which had primary responsibility for monitoring the fallout from nuclear tests in the Pacific.) He reported his results for the first time in 1960 and offered several preliminary findings. The instrument at La Jolla had been too influenced by random local variations in gas concentrations to be reliable, leading Keeling to conclude that such monitoring would need to be conducted at remote locations far from human activities involving combustion. The data from Antarctica showed little seasonal variation but displayed a suggestively rising trend in concentrations of CO_2 that seemed to confirm Revelle's hypothesis about fossil carbon being returned to the atmosphere.

Even more interesting were the measurements at Mauna Loa, which showed a marked seasonal rise and fall in CO_2 concentrations, with the highest levels being recorded in winter and the lowest in summer. (This same pattern appeared in the Arctic data as well.) Keeling soon realized that his instruments were literally recording the collective breathing of all plants and animals on the planet. Because land and therefore vegetation are concentrated in the Northern Hemisphere, summertime photosynthesis reduced CO_2 levels that rose again in winter, when plant respiration was at a minimum. Beneath these seasonal cycles, though, the Mauna Loa data showed the same subtle trend line as did those from the Arctic and Antarctic. The aggregate concentration of atmospheric carbon dioxide was gradually rising, an effect that was visible even with just two and a half years of measurements. If you turn back to the title page of this book and look closely at the squiggly line that rises across the two pages, the first three peaks on the lower left represent the period between 1957 and 1960 that Keeling reported in March of the later year. Although there were a few months in 1964 when he ran out of funding and it looked as if the monitoring might cease, it soon began again and has continued ever since. Both the seasonal oscillations and the rising trend line have continued unabated, with the latter now typically indicated by a seven-year moving average that eliminates seasonal cycling to show that Roger Revelle's geophysical experiment of fossil carbon returning to the biosphere is indeed well underway.

That is the origin story of the famous Keeling Curve that lies at the heart of Joshua Howe's *Behind the Curve: Science and the Politics of Global Warming*, one of the most important books yet written about the history of anthropogenic climate change since World War II. Howe seeks to trace not just the history of climate science, but also the ways in which science has interacted with politics as the rising trend of Keeling's curve has attracted growing public attention. At the heart of this book is a paradox. How can it be, its author asks, during a period when human understanding of climate dynamics has become ever more sophisticated, when we have never had more data or better models for analyzing the intricate interactions of earth, ocean, and atmosphere, that disagreements about climate policy have never been more intractable? How can our science be so good and our policy response so incommensurate to the scale of the climate threats that scientists have identified? Part of the answer, of course, is a well-funded and carefully orchestrated disinformation campaign, especially since 2006, to foster public skepticism about climate science. That campaign provides a kind of coda for *Behind the Curve*, but it is not the primary focus of this book. Howe's chief argument is more interesting and more troubling: the general ineffectiveness of climate scientists in advocating for public policies to address climate change is the vice of their virtue. The very qualities that have enabled them to produce good science have rendered them surprisingly ineffective as public advocates.

How can that be? Howe traces in fascinating detail several crucial tensions between climate science and climate politics that were already evident long before the political controversies that exploded in the early twenty-first century. One is the strange way in which climate change is arguably the largest but least visible of all environmental impacts that humanity has promulgated on planet Earth. Because its effects are so widely distributed, so gradual in their build-up, so subtle in their expression, and so unpredictable in their day-to-day occurrence, it is very hard for human beings to know with confidence whether they are or are not actually experiencing "climate change." Our most compelling evidence is based on data like the ones that the descendants of David Keeling's instruments still collect on Mauna Loa, observing minute changes in gaseous concentrations that are themselves undetectable to the human senses. Such data are then fed into ever more intricate computer programs that model the interactions among myriad variables to estimate future trend lines and the likely consequences of different concentrations of greenhouse gases. Such models have dramatically improved over time and are in fact our best available tools for understanding the effects of Roger Revelle's "large scale geophysical experiment"—but that is very far from saying that ordinary people understand how they work or feel confident about trusting them.

It is here that science and politics come into stark tension with each other. The scientists gathered the data, built the models, and thus made themselves gatekeepers and interpreters for the knowledge they created. Because what they have learned has persuaded them that grave threats lie on the horizon if greenhouse gas accumulations are not reversed, they have sought to share their knowledge with the public in the hope that doing so would encourage policy responses capable of mitigating the coming harms. (They have also shared this knowledge to encourage more funding for climate science research, a potential conflict of interest that became a weapon in the hands of adversaries who accused them of exaggerating claims to support their own work.)

But the scientists were at a disadvantage in two important ways. First, their own professional practice and statistical expertise required them to pay exquisitely close attention to the error bars surrounding their own instrumental data and model estimates, so that no conclusion could ever be stated with 100 percent certainty. Causality in the world of large-scale atmospheric movement is never absolute, and scientists are rarely comfortable with legal and political arguments that require certainty "beyond the shadow of a doubt." Science is at its best when it doubts and tests everything, whereas policy debates more often than not seek reliable assurances that can be asserted with near-total conviction. This made the language of science vulnerable to polemical attacks that turned its own greatest strength—its honesty about uncertainty and its perennial openness to critique—into a rhetorical weakness, especially for those who were already deeply suspicious of climate policy prescriptions that seemed to entail increases in government regulation, reductions in liberty and economic growth, and relentlessly secular world views.

The rhetorical problems of probabilistic causality were coupled with a second disadvantage that was if anything even more problematic: a deeply held scientific conviction that the solution to scientific uncertainty is more scientific research and ever more precise scientific models. Scientists had trouble recognizing the limits of scientific reasoning itself when they found themselves involved in public discussions that involved human values, cultural traditions, and political economic institutions that were resistant to rapid change. Although some of these "drivers" of climate change were eventually added to the models, it was generally as black boxes whose inner workings were only dimly understood by the model builders themselves. And because these black boxes did such an inadequate job of understanding or interpreting climate change in its larger human contexts, they have done a consistently poor job of influencing policy debates in the ways the model-builders hoped they might.

Here, then, is the challenge Joshua Howe puts before us in *Behind the Curve*, the title of which is a pun that nicely captures the most important elements of his argument. It is the scientists who understand the geophysical processes behind the rising curve of greenhouse gases that David Keeling began detecting in 1957, and it is the scientists who have framed our collective understanding of climate change. But because they have conceived of climate change mainly in scientific terms, they have had trouble recognizing the complex causal relationships within human societies that can never be adequately understood in the reductive mechanistic language of climate "drivers." As a result, public discussions of the human values at stake in climate change have yet to catch up with the sophistication of the scientific models. They are well behind the curve, terrifyingly so. This has left science open to cynical caricatures from "skeptics" with powerful self-interested reasons for encouraging as much uncertainty and obscurantism as possible. It has also left the rest of us with deeply inadequate tools for holding the very difficult conversations that are long overdue. Joshua Howe's conviction is that we must look beyond science for solutions to questions of human value that science alone can never answer. Only by placing climate change in a larger cultural and historical frame—as *Behind the Curve* consistently succeeds in doing—will we learn what we must from science without evading the ethical, moral, and political work that is no less essential if we are to find our way through the challenging choices that lie ahead.

ACKNOWLEDGMENTS

THIS BOOK BEGAN IN THE BACK ROW OF A SMALL LECTURE HALL AT Stanford University during a three-hour talk about sixteenth-century numismatics that almost made me quit graduate school. I was taking Jessica Riskin's graduate seminar, "The Construction of Scientific Facts," and rather than quit I began to think about how I could marry my interest in history with important contemporary issues in the context of her course. The search for something more important than coin collecting led me to Stephen Schneider, who introduced me to global warming and gave me unrestricted access to the four decades of climate change history stored in his memory and filing cabinets. That is when I began writing about the history of global warming, and that is also when I began accruing debts that I will never be able to fully repay. The first of these debts is to Steve, who passed away in 2010, and to his wife, Terry Root, for their patience, kindness, encouragement, and knowledge.

This book owes its greatest intellectual debt to Richard White. Anyone who knows Richard or his scholarship will see his fingerprints all over the best parts of this work. He is a brilliant teacher, a ruthless critic, a gentle mentor, and a loyal friend. I have benefited from each of these in equal measure. Thank you, Richard, for teaching me to love the work.

A great number of friends and colleagues have inspired my thinking about the history of global warming over the years, and it has been a pleasure and honor to work with each of them. Michael Reidy took a chance on me and gave me a job in a beautiful place and then pushed me to think deeply and differently about some of the things that I thought I knew about science and its history. He and Brett Walker have been important intellectual influences and also some of my strongest advocates, and they deserve a good deal of credit for bringing this book to fruition. Jessica Riskin nurtured this project from its infancy and has returned to it periodically when

I have needed her help the most. Beth Lew-Williams, Timothy Tomlinson, Lori Flores, and Kevin Kim all read early drafts of the manuscript, and their thoughtful critical feedback helped immensely in the formulation of this project. I am also indebted to Robert Proctor, Paula Findlen, Naomi Oreskes, Erik Conway, Bernie Lightman, Jacob Hamblin, Michael Egan, Kathy Morse, Mark Fiege, Spencer Weart, Gus Speth, Denis Hayes, Mickey Glantz, Thomas Malone, Chris Koski, Noelwah Netusil, Julie Fry, Sarah Schaack, Chris Wilkins, Peder Roberts, Jeff Miner, Andrew Gerhart, Lydia Barnett, Brett Burkhardt, Lindsay Hine, Craig Hine, Alden Woodrow, Nathaniel Shoaff, Bailey McCallum, and the Harnden family, as well as to the incredible groups of students at Stanford, Montana State, and Reed College who have pushed and energized me in different and unexpected ways.

I received quite a bit of help on the research for this book, beginning with the friendly, patient, and impossibly knowledgeable archivists at the National Center for Atmospheric Research (NCAR), the Scripps Institution of Oceanography, the American Association for the Advancement of Science, the Bancroft Library, the University of Colorado at Boulder Special Collections, the Ronald Reagan Presidential Library, and the National Archives. Special thanks to Diane Rabson, Lynda Claussen, Norma Rosado-Blake, and Amy Crumpton. The graduate program in the Department of History at Stanford University provided support and an academic home through much of my research. The department's Lane Grant, alongside NCAR's Center for Capacity Building, provided funding for early archival research. I also had the privilege of working closely with scholars in Stanford's Interdisciplinary Program in Environment and Resources, the Woods Institute for the Environment, and the Bill Lane Center for the American West. The American Meteorological Society's History of Science Fellowship and the G. J. Lieberman Fellowship at the Stanford Humanities Center supported this work through the end of my graduate career. More recently, I received warm support and encouragement from the faculty and staff in the Department of History, Philosophy, and Religion at Montana State University and in the Departments of History and Environmental Studies at Reed College.

William Cronon has provided predictably brilliant feedback on this work since we first began to talk about this as a book in the Weyerhaeuser Environmental Books series. I count myself fortunate to have had the benefit of his great creative intensity in taking this project from dissertation to manuscript. Marianne Keddington-Lang has served as a point person, guide, and interpreter throughout the publishing process. The quality of the final product is a credit to her and her colleagues at the University of Washington Press.

Finally, it gives me immense pleasure to thank my loving and irrepressible parents, Amy and Steve, and my astoundingly creative and intelligent sisters, Jessica and Allison, for seeing me through the ups and downs of life and work over the past decade. Thanks, too, to Jim and to my kind and remarkable friends who have, over the years, become another layer of family. This book is for my uncle, Lee Schipper (1947–2010), and for my nephew Graham.

ABBREVIATIONS

AAAS	American Association for the Advancement of Science
AEC	Atomic Energy Commission
DOD	Department of Defense
DOE	Department of Energy
ECOSOC	U.N. Economic and Social Council
EPA	Environmental Protection Agency
FAO	U.N. Food and Agriculture Organization
ICSU	International Council of Scientific Unions
IGY	International Geophysical Year of 1957–58
IMF	International Monetary Fund
IPCC	Intergovernmental Panel on Climate Change
ONR	Office of Naval Research
OSTP	Office of Science and Technology Policy
NAS	National Academy of Sciences
NCAR	National Center for Atmospheric Research
NEPA	National Environmental Policy Act (1969)
NOAA	National Oceanic and Atmospheric Administration
NRC	National Research Council
NRDC	Natural Resources Defense Council
NSF	National Science Foundation
ppm	parts per million
PSAC	President's Science Advisory Committee
SCEP	*Study of Critical Environmental Problems* (1970)
SCOPE	Scientific Committee on Environmental Problems (of the ICSU)
SMIC	*Study of Man's Impact on Climate* (1971)
UNCED	U.N. Conference on Environment and Development (Rio de Janeiro, 1992); also known as the Rio Earth Summit

UNDP	U.N. Development Programme
UNEP	U.N. Environment Programme
UNFCCC	U.N. Framework Convention on Climate Change
USACDA	U.S. Arms Control and Disarmament Agency
USAID	U.S. Agency for International Development
USIA	U.S. Information Agency
WMO	World Meteorological Organization

BEHIND THE CURVE

INTRODUCTION

Telling Stories about CO$_2$

FEW PEOPLE WHO DO NOT WORK IN CLIMATE SCIENCE OR GLOBAL warming advocacy recognize the Keeling Curve by name. Even for the well-informed, the name typically rings only a faint and distant bell. And yet, when you actually see the undulating, upward-sloping line running from left to right, between x and y axes, in the context of a discussion about global warming, the image and its meaning become instantly familiar. The Keeling Curve represents the measured concentration of carbon dioxide in the atmosphere at the Mauna Loa Observatory since 1958, when Charles David Keeling began measuring CO$_2$ with his manometers there.[1] The undulations represent annual cycles of growth and decay on the land masses of the Northern Hemisphere, where plants fix carbon in leaves and stems in the spring and summer months and then release carbon back into the atmosphere as they drop those leaves in the autumn and winter. In 1958, Keeling found that for every million units, or "parts," of the mélange of gases we call the atmosphere, 315 were CO$_2$. In 2013, that number—parts per million CO$_2$—surpassed 400.[2]

The information needed to interpret the significance of the Keeling Curve is neither particularly complicated nor particularly new. Between 1859 and 1862, the Irish physicist and mountaineer John Tyndall used a new instrument called a ratio photospectrometer to confirm experimentally that CO$_2$ absorbed radiation and therefore acted to regulate the temperature of the atmosphere and the earth.[3] In 1896, a diligent Swedish chemist and Nobel laureate named Svante Arrhenius did a set of tedious pencil-and-paper calculations that showed that a doubling of CO_2 would cause about a 5°C increase in the mean temperature of the earth.[4] And in 1957, just before Keeling's project took off, his oceanographer boss, Roger Revelle, and a radiochemist named Hans Suess demonstrated geochemically that humans were, as a few scientists before them had claimed, contributing CO_2 from fossil fuels to the global atmosphere. Of course, each of these ideas has undergone significant modifications in the past half century as our understanding of the dynamics of the global atmosphere has grown. But I am only half joking when I tell my students that in order to understand the significance of the Keeling Curve, they need little more than the physics of the 1860s, the mathematics of the 1890s, and the chemistry of the 1950s. Equipped with that information, they will find the Keeling Curve to be one of the simplest and most powerful images in the iconography of anthropogenic climate change.

As a visualization of data, the Keeling Curve derives its power from its simplicity. The curve measures a single variable—atmospheric CO_2,—over time. But the simplicity of the Keeling Curve masks much more about CO_2-induced climate change than it reveals. Behind the curve lies a series of contextually specific interactions between individuals, institutions, ideas, and interests. Behind the curve lies the growth of a fossil fuel–dependent industrialized world producing CO_2 and the long hours spent by scientists and institutions measuring, describing, and predicting the causes and effects of the rise in atmospheric CO_2. Behind the curve rests the passion of a community of concerned scientists and citizens interested in mitigating the climatic and environmental changes caused by rising CO_2. And behind the curve also sits the interactions between a large cast of local, national, and international political actors and institutions charged with creating a framework for regulating CO_2 and other greenhouse gases. The curve remains relatively constant in its shape and slope, but with every annual oscillation, the constellation of forces behind the curve has changed. To divorce the Keeling Curve from the complexities that it masks and the contexts in which it appears is to miss what makes the image both so incredibly powerful and so puzzlingly problematic. The Keeling Curve seems simple and elegant, but the problem that it describes is not simple at all.

This book traces the scientific and political history of climate change over the life of the Keeling Curve. It is a story about the scientific, political, environmental, and economic complexities that the simple measure of atmospheric CO_2 since 1958 tends to hide and how those complexities have changed over time. The Keeling Curve is only one in a variety of simplified descriptions of the earth's climatic and environmental systems; but like the Keeling Curve, each of these models masks as much as it reveals about the interactions between humans and the global environment. This story contains little that is new about the technical details of these models, but it contains much that is new about the people, communities, and institutions that have created these models, about how these models have been interpreted in different historical and political contexts, and ultimately about why both these models and the people who have created and interpreted them have yet to find success in the battle to combat climate change.[5] This book is about what lies behind the curve.

This point about failure represents the first riddle of the Keeling Curve: the central problem of the political history of global warming. Keeling's initial measurements of atmospheric CO_2 at Mauna Loa and later in Antarctica marked the beginning of a half century of scientific inquiry into the place of CO_2 in the processes of the global atmosphere. The issue has become one of the major social and political challenges of the twenty-first century. Since 1958, scientists have made great strides in measuring, modeling, explaining, and predicting the increase of CO_2 and its potential impacts on the global environment. They have introduced the potential dangers of anthropogenic climate change to both the American and the global public, and alongside environmentalists they have created national and international scientific and political frameworks for mitigating CO_2-induced warming and the climatic disruptions associated with it. If it were possible to distill these developments into simple measurements of "levels of knowledge" and "levels of political investment," you could imagine the graphs of these variables over time running roughly parallel to the Keeling Curve. Perhaps these graphs would be less linear: they might cross the measure of CO_2 at critical junctures as knowledge and interest spike in periods of apparently geometrical growth and then plateau during periods of relative stasis, even decline. In the long run, however, these curves of knowledge and political investment would show tremendous growth. As a society, we know and care much more about CO_2 and climate change than we did in 1958.

And yet, after more than fifty years of scientific inquiry and political activism in the service of reducing humans' dependence on fossil fuels and lowering CO_2 emissions, the upward trajectory of the Keeling Curve has held steady. Fossil fuel energy continues to support an expanding global economy, and greenhouse gas emissions consequently continue to expand as well. As a result, the world is getting warmer. In fact, not only does the Keeling Curve show a consistent increase in the concentration of CO_2 in the atmosphere, it also shows a rise in the rate of increase, from about 1 ppm (part per million) per year to closer to 2 ppm per year.[6] The problem is not getting better; it is getting worse. To spin the title of this book another way, after fifty years we are still behind the curve. This book is about why this is so.

THE ARGUMENT

It is tempting to answer questions about why society has failed to deal with climate change by pointing at government officials, skeptical scientists, oil industry lobbyists and executives, conservative politicians, and financial institutions that have opposed and often undermined scientific research and congressional legislation since the issue first appeared on the public radar.[7] The opposition is certainly out there, and as Naomi Oreskes and Erik Conway have shown in their study of the contrarian scientists involved in setting climate change policy for the Ronald Reagan and George H. W. Bush administrations, their influence has been real and pernicious.[8]

But powerful and vocal as opponents of global warming policy have been since 1990, this political opposition has mostly *reacted* to the history of climate change. From the beginning, it has been the groups of people most interested in studying and mitigating CO_2—climate scientists, environmentalists, and a few politicians, government bureaucrats, and diplomats—who have set the terms of the debate. To borrow and retool a model of peasant resistance proposed by James C. Scott, in the history of global warming the structures of scientific research and political advocacy have helped define the structures of scientific and political resistance.[9] Our collective failure on global warming has as much to do with the ways that scientists and environmentalists have approached and presented the problem as with how their opponents have resisted possible solutions.

Our collective social and political failures on global warming stem in part from the primary and sometimes exclusive focus on science in global warming advocacy. Since anthropogenic climate change first emerged as a potential environmental problem in the twentieth century, scientists, environmentalists, and politicians have consistently prioritized scientific information over social or political change. As a result of their science-first approach, scientists have made tremendous strides in their understanding of CO_2 and climate change. But this narrow focus has shaped a broader political conversation that is also primarily scientific, and here the primacy of science has at times forestalled meaningful political and moral engagement with the problem. To the extent that climate change discourse continues to revolve around science and has failed to generate real political change, advocates' science-first approach has actually undermined many of their objectives. In narrative terms, their story fits the familiar form of a tragedy.[10] This is a big claim, and it is important to begin with a sketch of how the tragedy—and the argument—works.

The tragedy of global warming begins with the intangible nature of the problem itself. The Keeling Curve shows that CO_2 is rising, but you cannot see CO_2. You cannot hear how 43 percent of the CO_2 released by the burning of fossil fuels mixes with and resides in the global atmosphere, and you cannot touch the way in which that extra CO_2 absorbs and reradiates energy from the sun.[11] You cannot smell the subsequent increase in global mean temperature; and though you may be able to taste the Valencia oranges that will grow in Oregon two decades from now, their sweetness and geographical novelty can taste like global warming only if you associate those oranges with processes that occur on geographical and temporal scales that are unavailable to your senses.[12] Twenty-first-century Inuit and Micronesians whose lands and livelihoods are disappearing due to melting sea ice and sea-level rise will confirm the reality and immediacy of the impacts of anthropogenic climate change, but these impacts have not been apparent throughout the majority of the history of global warming.[13] Even now, understanding these environmental challenges requires understanding the global processes that transcend our immediate relationship with the world.

That is not to say that we cannot or do not engage intellectually with global processes and global spaces. We do. But when we encounter climatic change in terms of the processes of the global atmosphere, our experience is almost always mediated by science. Historically, only scientists have had the expertise, the technologies, and the language to understand and communicate the phenomena of this global space. To the extent that environmentalists, politicians, and the public understand global warming and the global atmosphere in these issues' geographical totalities, they understand them in the language of science. To paraphrase climate science historian Paul Edwards, scientific models are the only way any of us really know anything about global warming.[14]

Understanding climate change has not just been a problem of description and prediction, however. As the impacts of increased CO_2 on global temperature began to appear on scientists' radar in the late 1950s and early 1960s, so too did it become clear that mitigating these impacts would require social and political change. When scientists began to realize in the 1960s that changes in the global atmosphere wrought by increasing CO_2 and other human activities could threaten environmental and human systems, their exclusive access to these global processes and spaces forced them to serve as climatic spokespeople. Very early on in this story, they became more than gatekeepers; they became advocates.

This has been problematic. Since the 1960s, climate scientists have approached their roles as advocates uneasily. Science derives its authority in large part from its claims to community-defined standards of objectivity, and often those standards presuppose political neutrality.[15] As a group, scientists are by professional training and often by personality predisposed to avoid political advocacy. In the case of climate change, the particularities of the problem forced the very group of people most averse to marrying professional and political goals into positions of advocacy that required them to do just that. Their ambivalence about climate change advocacy appears as a theme throughout the history of global warming. After the intangibility of the problem and scientists' unique access to it, this ambivalence is the third key component in understanding this history in the narrative terms of tragedy.

Each iteration of climate scientists' advocacy conundrum in this story looks different, but their response to the scientist-as-advocate paradox has followed a trend: they have advocated for more and better science. More and better science in the hands of the right political actors, they argued, would inevitably lead to responsible political action on climate change. "More and better science" meant different things to different people in different contexts over time, but again and again climate change advocates—scientists and otherwise—espoused a belief in what one congressional science advisor in the 1970s called the "forcing function of knowledge," whereby a better scientific understanding of the problem of climate change would force appropriate political action.[16] If the history of global warming unfolds as a familiar tragedy, then this overweening faith in the power of science to inspire political action—the commitment to the "forcing function of knowledge"—is what drives the failure of climate change advocacy, our heroes' collective tragic flaw.

There are two important things to understand about the problematic primacy of science in global warming advocacy. First, naïve as a belief in the "forcing function of knowledge" may seem, the top-down, science-first approach to climate change advocacy initially granted scientists remarkable access to both resources and decision making at the national and international levels. Climate scientists' influence both within the U.S. federal bureaucracy and in the United Nations helped put CO_2 and climate change on the geopolitical map. Second, that same top-down, science-first approach left climate change advocates vulnerable to political change. The problem was not only that science-first advocacy stopped working as political conservatives and their industry allies came to power in the 1980s, as Spencer Weart and others have shown.[17] Scientists' and environmentalists' reliance on science as a driver of political action also gave their opponents a clear target: science itself. Soon the same scientists who argued that the forcing function of knowledge would drive better policy found themselves engaged in a political struggle over the control of knowledge. As a consequence, there has been much more debate about what the Keeling Curve means than there has been real change in the social and political factors that determine its slope.

It is worth pausing to reflect on the key mechanism of this tragedy, because Shakespearean tragedy has conditioned us to interpret the "tragic flaw" of narrative in a way that I think can be misleading. For Shakespeare's tragic heroes—Julius Caesar, Macbeth, King Lear—*hamartia* is a fundamental flaw of character, such as hubris, ambition, greed, or vanity, often manifest in a decision or set of decisions that leads to the main character's undoing. But the original Greek term *hamartia,* the defining element of Aristotle's "tragedy," is more nuanced. The word comes from the world of archery, a "missing of the mark"—not quite as simple as a mistake but neither as damning as a Shakespearean vice.[18] When it comes to the tragedy of global warming, it is an important distinction. There are no Shakespearean characters in this story, no true heroes and no true villains. In this story, *hamartia* is not Shakespearean. When I say that climate change advocates' overweening faith in science is the tragic flaw in this narrative, I envision an archer whose confidence in her equipment causes her to overlook a faulty sight, frustrating even her best and most determined efforts as every shot flies high and to the right.

THE STORY AND ITS STRUCTURES

Tragedy provides a good model for understanding the shortcomings of climate change advocacy from the early 1950s until the beginning of the twenty-first century. The main characters undermine their ultimate objectives through the very means employed to achieve those objectives, and in that sense the history of climate change fits Alfred North Whitehead's definition of a narrative tragedy. The story is tragic in a colloquial sense as well. The failure of climate change advocates to overcome the limitations of science-first advocacy has had real and destructive consequences for individual organisms, species, ecosystems, and human populations, and these consequences can rightly be called tragic.

But tragedy is not the only narrative element at work here. It also might be helpful to think in terms of a less traditional form of narrative. It might be helpful to think in terms of memes.

In his 1976 book, *The Selfish Gene*, Richard Dawkins described memes as historically specific cultural artifacts analogous to genes in the ways they are replicated, inherited, and reinterpreted from generation to generation.[19] The Keeling Curve—and the idea of rising atmospheric CO_2—behaves like a meme in the history of global warming, gaining new meaning as it is passed from one cultural context to the next. With each iteration of the meme since its inception in 1958, each reinterpretation of the Keeling Curve in a new historical context, the curve has taken on new meanings to new groups of people with different interests and agendas. Each iteration both inherits and modifies the nuance and complexity of the last. The problem of climate change becomes more complex as you trace the curve forward in time and as the meme acquires more layers of interpretation.

The Keeling Curve can help illustrate how these iterative stories fit into the larger picture. The most common representation of the Keeling

Curve contains two lines: a smooth, upward-sloping line of seasonally corrected data that approximates the annual average atmospheric CO_2 as measured at Mauna Loa since 1958 (typically in black) and an undulating line of monthly mean atmospheric CO_2 that oscillates above and below the annual average (often in red). Behind both lines on the Keeling Curve lie individual measurements of CO_2, the raw data from which scientists at the Scripps Institution of Oceanography and the National Oceanic and Atmospheric Administration construct the two lines on the graph. The lines of monthly and annual averages help give the raw data meaning by presenting them in two different but related ways. Similarly, in this book I present the history of climate change in two related ways: one a single, overarching tragedy, the other a set of contextually specific stories that taken together make up the substance of that tragedy. Behind my two stories lies a body of evidence analogous to the raw data of the Keeling Curve, the documents and interviews upon which my narrative is based. The coarse-grained tragedy of global warming and the detailed, contextually specific stories it comprises represent two ways of making meaning of that evidence. In the Keeling Curve, the undulating line repeats a story of annual oscillation—a repeating tale much like the historical reinterpretations of CO_2 that make up this book. Aggregated and simplified into a coarser average, these repeated annual cycles—and these iterations of the meaning of increasing atmospheric CO_2—describe the smooth, solid line of annually averaged CO_2 over time, a line representative of tragedy.

THEMES AND REFRAINS

Iterative stories force us to look at a narrative multiple times and think about what has changed and what has remained the same. This approach not only reveals the central refrains of the story—the intractability of global warming and the problematic primacy of science in our attempts to solve it—but also highlights a number of secondary narratives that give those refrains texture and historical meaning.

The first and perhaps most striking of these secondary themes is the importance of the Cold War in shaping both the science and politics of climate change through the second half of the twentieth century. Several historians have explored how American scientists capitalized on an expanding Cold War research system in the 1950s to advance the study of

CO_2 and climate change.[20] Especially following the Soviet Union's launch of *Sputnik* in 1957, entrepreneurial scientists and science administrators like Roger Revelle found that linking CO_2 to a growing Cold War preoccupation with geophysical processes helped tie atmospheric scientists to the sources of government money that would support climate change research. The Keeling Curve was a product of that Cold War scientific entrepreneurship. Funding for Keeling's measurements of CO_2 came from the Atomic Energy Commission—perhaps the quintessential Cold War agency—in a grant slated for research associated with the International Geophysical Year of 1957–58 and funneled through the Scripps Institution of Oceanography by Roger Revelle, the self-described "granddaddy of global warming."[21]

But the relationship between the Cold War and climate change entails much more than climate scientists' stake in the federal science budget. Scientists and their colleagues at environmental organizations and in Congress wove Cold War hopes and Cold War anxieties into the fabric of climate science and global warming politics. The Cold War lurks behind nearly every iteration of the CO_2 story in the last half century, and it crops up frequently, often in unexpected ways. Perhaps the most notable and heretofore unexplored influence of the ideological conflict on global warming politics lies in the Cold War's rapid and unexpected end. Between 1988 and 1992, scientists and diplomats hashed out the details of the two most important institutions of international climate change politics—the Intergovernmental Panel on Climate Change and the U.N. Framework Convention on Climate Change—quite literally in the midst of the Cold War's final years. Both institutions reflect the then-changing geopolitical milieu. The Cold War not only helped define the historical contexts in which scientists, environmentalists, and politicians interpreted CO_2; the ghost of the Cold War continues to influence the way we understand CO_2 and climate change today.

The influence of the Cold War entwines with a second theme that reappears throughout the stories of CO_2 and climate change in the last half century: the fraught relationship between climate scientists and America's professional environmental organizations. From the beginning, scientists have not only struggled with their ambivalence about serving as advocates; they also have struggled with their association with environmentalism. Specific moments in the history of climate change science and advocacy have forced both groups to reinterpret their relationship to each other and to the issue of CO_2-induced climate change. It was not until the 1980s that the two groups came together in any kind of permanent way, and their marriage was largely a by-product of another reconsideration of their relationship forced on them by changing historical circumstances after the election of Ronald Reagan. By the end of the century, climate change had become a central concern of local, national, and global environmental organizations—so much so that it is easy to overlook the extent to which a relatively small group of people actively made global warming the "environmental" issue that we now know.

THE WORDS

The Keeling Curve has been interpreted not only in contextually specific ways but also with an evolving set of terms. The basic physical phenomena at the heart of the story can be difficult to pin down in language. Scientists and politicians have used "climate change," "global climate change," "CO_2-induced climate change," "anthropogenic climate change," "global warming," "anthropogenic global warming," "greenhouse warming," and any number of variations on these general themes to describe different aspects of climatic and atmospheric change. I find Thomas Friedman's concept of "global weirding" particularly useful in relating the problems of climatic variability that are most disconcerting about global warming. The problem, Friedman explains, is not that greenhouse gas emissions raise the average temperature of the earth. The problem is that behind the measure of "average" temperature is a host of unpredictable changes in local and regional weather and climate patterns for which neither ecosystems nor human populations are prepared. Record Sierra snowfalls amid decades of drought are not immediately problems of temperature; rather, the global increase in temperature makes local weather and climate difficult to predict or plan for. It makes the climate "weird."[22]

Friedman's is a good idea, but I fear that adopting his language would only add to the verbal confusion of an ever-expanding climatic lexicon. I have tried to simplify by sticking with the two most lasting and ubiquitous terms associated with atmospheric CO_2: *climate change* and *global warming*. I follow Mike Hulme's succinct and helpful distinction here.[23] *Climate change* refers to past, present, or future changes in climate, regardless of direction (warmer or cooler) or cause ("natural" or anthropogenic). It is currently the preferred nomenclature of scientists studying the physical phenomena of the global atmosphere and is the term employed by the Intergovernmental Panel on Climate Change. *Global warming* refers to a specific breed of climate change characterized by an increase in global mean temperature caused by human-induced increases in atmospheric greenhouse gas concentrations. It is a historically specific term that arose in the late 1970s, and since 1980 it has carried with it a carload of political baggage that is, to a large extent, the subject of this book.[24] To put it glibly, global warming is human-caused climate change, plus politics.

The suggestive etymology and history of the term *climate* bear reflection as well. Derived from the Greek *klima*, or "inclination"—as in, inclination of the sun from the earth—*climate* denotes the average weather of a place over time.[25] The concept reflects an awareness of the relationship between latitude (which correlates with the inclination of the sun) and the "normal" weather conditions of a place over the course of months, years, and decades. But as Lucien Boia chronicles in *The Weather in the Imagination*, describing and explaining climatic conditions has at once reflected and supported societies' beliefs about other cultures.[26] From Pythagoras's disciple Parmenides and his five climatic zones to the early twentieth-century climatic determinism of Ellsworth Huntington, when scientists and philosophers of the past have written about climate, they have typically really been writing about people. It is perhaps not surprising, given this long history, that in the twenty-first century it has proven difficult to separate the science of climate from its politics.

Finally, a brief note about gases. CO_2 is not the only greenhouse gas, not the only substance that causes climate change. In fact, on a per molecule basis, methane (CH_4), ground-level ozone (O_3), oxides of nitrogen (NO_x), and other gases warm the atmosphere orders of magnitude more powerfully than CO_2.[27] I focus on CO_2 because until recently the history of global warming has been primarily the history of the CO_2 problem. Even today, despite carbon dioxide's relative lack of warming power, its connection with the fossil fuels at the center of the global economy makes it the most important—and difficult—greenhouse gas to mitigate. Its persistence as a key unit of climate science underscores its centrality. The warming potential of other greenhouse gases is measured in relation to the warming produced by one unit of CO_2.

THE PURPOSE OF THIS BOOK

The last thing to know before you dig into the first chapter is what this book is for. First, let me tell you what I am not trying to do. I am not trying to convince you that global warming is real, though I am convinced by the evidence that it is. I probably would not have written this book otherwise. Neither am I trying to sell you on a particular solution to the problem. I do not have one, nor do I think there is a single solution. I have not written an all-inclusive, up-to-the-minute history of global warming, and I am not going to claim that this book contains everything you need to know about the science and politics of global warming. It does not.

What this book does contain is a thoughtful mixture of narrative and analysis that highlights how scientists, environmentalists, and the broader public have interpreted the problem of CO_2-induced climate change in different contexts over time. These stories demonstrate the patterns of interpretation and advocacy that have led to failure in the past in order to make it easier to recognize similarly problematic patterns in the future. This book gives you the tools to interpret the next iteration of the story of CO_2, to recognize what has stayed the same in the climate change conversation and to assess the value and efficacy of what has changed. In short, this is a history for the future.

1

THE COLD WAR ROOTS
OF GLOBAL WARMING

IN THE ARCHIVES OF THE SCRIPPS INSTITUTION OF OCEANOGRAPHY IS a photograph of a young Roger Revelle—age twenty-seven or so—sorting specimens on the deck of the collecting vessel *E. W. Scripps*. The year is circa 1936, and Revelle's setup looks decidedly ad hoc. He sits on what look to be the wooden slats of the aft deck, right leg bent beneath him and left leg splayed out, amid an assortment of mason jars, collecting rags, and a bucket. In a black wool sailor's jacket, with one sleeve rolled up and cuffed white trousers stained at the knees, he is hardly the picture of a careful, plodding scientist. He looks a little bit like I imagine James Dean might have, had he played an oceanographer: intensely focused and unflappably cool, taking care of the business of collecting specimens while enjoying the ocean breeze on a boat named for a member of the family into which he had married—the namesake also of the Scripps Institution of Oceanography, from which he would earn a Ph.D. and which he would eventually direct. More than anything, as Revelle looms over the little mason jars, his hand enveloping whatever tiny specimen he is picking up at the moment, the man in the photograph—the self-appointed "'granddaddy' of the theory of global warming"—looks altogether too big for whatever specific form of science he is conducting.[1]

Roger Revelle was a good oceanographer. He was an excellent science administrator, and he was an even better science advocate. He was also interested in CO_2, and perhaps no individual in the 1950s and 1960s did more than Revelle to put atmospheric CO_2 and climate change on the Cold War research agenda.[2] In 1958, when Charles David Keeling began permanent operations at Mauna Loa and atmospheric CO_2 measured about 315 ppm, Revelle was Keeling's boss.

The introduction of CO_2 into the Cold War research system marks the beginning of the political history of global warming. Between 1955 and 1963, Revelle and a handful of creative and influential scientists capitalized on existing government research to gain funding and support for specific projects that involved measuring and monitoring atmospheric constituents like CO_2. They tapped into a pervasive interest in geophysical research that might have a bearing on weather modification, nuclear test detection, fallout, or other defense-related subjects in order to solicit funding and material support for atmospheric science. Soon, leaders in atmospheric science realized that in order to more fully study the processes of the atmosphere, they needed to secure funding for scientific institutions that could support long-term projects. Again, they capitalized on the potential security implications of geophysical research, and again they incorporated CO_2 into the heart of atmospheric science.

Tying the study of CO_2 to Cold War research enabled scientists to revisit CO_2 with new eyes, but there were two sides to the Cold War research coin. On one side, funding, new technologies, and institutional support gave scientists access to better data that confirmed that CO_2 had, in fact, begun to rise, and that its increase could have geophysical consequences. Revelle in particular framed CO_2 rise as a form of natural experiment, one that the new tools and technologies of Cold War science could help to monitor. On the other side of the coin, however, scientists harbored anxieties born of the Cold War, and these anxieties influenced how they structured their institutions, employed their new Cold War resources, and interpreted study results. As a result, in the 1950s and 1960s, research on atmospheric CO_2—and atmospheric science more broadly—reflected both the greatest hopes and the deepest fears of the Cold War milieu from which it sprang.

THE GRAND EXPERIMENT

Revelle's contributions to climate change history were both administrative and scientific. His most important scientific contribution involved a reevaluation of the role of the oceans in the global carbon cycle, a study supported by funds from the Office of Naval Research (ONR) and the Atomic Energy Commission (AEC).[3] In the early 1950s, nuclear weapons tests dumped large, known amounts of radiation into the atmosphere at specified times and places, and both the ONR and AEC had a keen interest in where that radiation went. One way to trace it was by following radioactive carbon through the atmosphere. In fact, alongside naturally occurring cosmic radiation, nuclear tests created a nearly ideal experiment for tracing the circulation of carbon through both the atmosphere and the oceans.[4] Amid nascent concerns about nuclear fallout from bomb tests at Bikini Atoll (where then-navy commander Revelle had headed a team studying a coral lagoon in 1946) and elsewhere, the government began to ask oceanographers like Revelle just how fast the oceans could swallow up the carbon that the U.S. military was irradiating in the atmosphere.[5]

In 1955, Revelle began working on the problem with Hans Suess, an Austrian-born physical chemist, nuclear physicist, and radiocarbon dating expert whom Revelle had hired at Scripps with ONR/AEC funding. While at the U.S. Geological Survey's laboratory in Washington, D.C., Suess had described a characteristic of the carbon cycle that would help Revelle in his attempts to use measurements of radiocarbon in air and seawater to investigate the rate of CO_2 exchange between the oceans and the atmosphere.

As plants grow, they "fix" the carbon from atmospheric CO_2 in their leaves, trunks, stems, and shoots, including a naturally occurring radioisotope of carbon called carbon 14 (C^{14}).[6] In 1949, Willard Libby of the University of Chicago devised a method for measuring the rate at which C^{14} decays, giving archeologists and paleontologists a new way of dating very old relics and the dirt they came from—a method called radiocarbon dating. Suess, interested in geochemical cycles, applied Libby's idea a little differently. Not all carbon is C^{14}; most, in fact, is "normal" carbon, or C^{12}. Using Libby's rate of decay, Suess calculated what the overall ratio of C^{14} to C^{12} in the atmosphere *should* be, given approximate time scales for major influences on CO_2 such as weathering and ocean mixing. The ratio came out low—that is, there was less C^{14} in the atmosphere relative to C^{12} than Suess's calculations predicted. He concluded that the whole mix must have been diluted, most likely by carbon from some ancient source in which most or all C^{14} isotopes would have decayed. Fossil fuels, which are stored in the earth's crust for millions of years and thus have little if any C^{14}, presented just such a source. The dilution of atmospheric carbon with fossil fuel carbon is known as the Suess effect. It provided a radiochemical confirmation of a contention made in 1938, by an obscure British steam engineer named Guy Stewart Callendar, that fossil fuel use contributed CO_2 to the atmosphere.[7]

At first, Revelle and Suess both assumed—like many other scientists— that the oceans would absorb this excess CO_2, and the two men hoped to study and trace radioactive CO_2 as it circulated through the atmosphere and found its way into the seas. Shortly before they sent their paper on the subject to the journal *Tellus* in 1957, however, it occurred to Revelle that they had failed to account for the tendency of seawater—a complex and often poorly stirred chemical mélange—to retain a generally constant acidity through a self-regulating "buffering" mechanism involving CO_2.[8] The CO_2 that most scientists assumed would be absorbed might just as easily be re-released while still at the ocean surface, resulting in the overall increase in atmospheric CO_2 that Callendar, for different reasons, had predicted. That increase could in turn have interesting and far-reaching geophysical significance.

"Human beings," Revelle famously wrote, "are now carrying out a large scale geophysical experiment of a kind that could not have happened in the past nor be reproduced in the future. Within centuries we are returning to the atmosphere and oceans the concentrated organic carbon stored in sedimentary rocks over hundreds of millions of years. This experiment, if adequately documented, may yield far-reaching insight into the processes determining weather and climate."[9]

In retrospect, Revelle's "grand experiment" statement has been framed as an early warning on global warming, but at the time Revelle and Suess were more curious and excited than anxious. As a colleague later remembered, "Roger wasn't alarmed at all... he liked great geophysical experiments."[10] Indeed, Revelle's realization about buffering was a late addition to the Revelle-Suess paper, which on the whole actually challenged the extent of the so-called Callendar Effect. Nevertheless, the paper's conclusion— and, more importantly, its authors—made an impact on the way scientists and their government sponsors interpreted atmospheric CO_2 for the next half century. That impact began with the advent of the Keeling Curve.

For Revelle, the immediate danger in the grand experiment was that it might not be adequately monitored and documented.[11] Master of the science-funding alphabet, he took steps to ensure that it was. Taking advantage of resources available to him as the president of the Scientific Committee on Ocean Research (SCOR), a committee of the International Council of Scientific Unions (ICSU), Revelle designed an atmospheric-monitoring program for the 1957–58 International Geophysical Year (IGY).[12] In July of 1956, the Scripps Institution of Oceanography had hired Keeling as a junior scientist, and he brought his pathological obsession with measuring atmospheric CO_2 with him to La Jolla. In 1957, Revelle set Keeling up with IGY funds (and about $10,000 of dubiously allocated AEC money) and put him in charge of the new IGY atmospheric-monitoring program.[13] Keeling constructed CO_2-monitoring stations at the Mauna Loa Observatory in Hawai'i and at a research post in Antarctica in order to establish a baseline of atmospheric CO_2 that could be used to measure future changes.[14] In March of 1958, the amount of CO_2 in the atmosphere stood at about 315 ppm.[15]

CO_2 AND THE INTERNATIONAL GEOPHYSICAL YEAR

The Keeling Curve would eventually become an icon of global warming, but in the spring of 1958 the curve was still only a few data points on a plotter, and Keeling was little more than "a peculiar guy" living on a vol-cano and trying to figure out the best way to measure CO_2.[16] Technically, Keeling's Mauna Loa Observatory put CO_2 research on the government payroll, but it was Revelle's involvement in planning the International Geophysical Year that really landed CO_2 on the public research agenda.

Lasting from July 1, 1957, to December 31, 1958, the IGY was the larg-est cooperative international scientific research effort the world had ever seen. The idea arose from a 1950 gathering of physicists interested in the ionosphere; in less than a decade it blossomed into a project involving more than sixty thousand scientists and technicians from sixty-six nations.[17] The initial group met at the home of James Van Allen, a rocket scientist concerned primarily with cosmic rays. Also present were Lloyd Berkner, a physicist who would become both president of the ICSU and a member of Eisenhower's Science Advisory Committee; Sidney Chapman, a man that *New York Times* science writer Walter Sullivan dubbed the world's "greatest living geophysicist"; and three other physicists, J. Wallace Joyce,

S. Fred Singer, and Ernest H. Vestine.[18] Envisioned as an updated version of the International Polar Years of 1882–83 and 1932–33, the IGY had a major polar science component but focused primarily on the physics of the earth's atmosphere. In fact, with projects in meteorology, climatology, ionospheric physics, aurora and airglow, cosmic rays, and solar activity, the IGY agenda defined what would become known more broadly as the field of "atmospheric science."[19]

Scientists designed the IGY to tackle a range of large-scale research interests, but from the beginning it was also a deeply political affair. Not surprisingly, in the international arena the politics of the IGY revolved primarily around Cold War U.S.-Soviet relations. These politics played out in the language of institutional acronyms.[20] The IGY was administered by the International Council of Scientific Unions (ICSU), with support from the United Nations Educational, Scientific, and Cultural Organization (UNESCO) and the World Meteorological Organization (WMO). Under Stalin, however, the Soviet Union had decided not to adhere to the ICSU, and that decision persisted through the process of "de-Stalinization" in the 1950s (the Soviets did belong to the International Astronomical Union and the WMO, however). The Soviet Union was thus not included in the original list of twenty-six invited nations, and the U.S. National Committee for the IGY had to persuade the ICSU to send the Soviets a special invitation.[21] The Soviets took eighteen months to accept, and once they did, their cooperation was limited. Soviet officials resisted proposals to allow scientifically oriented overflights by foreign nationals within Soviet borders, and Soviet scientists refused to discuss their nation's rocketry and artificial satellite programs. The rest of the Soviet IGY program was every bit as extensive as the United States' proposed plan—perhaps more so—but cooperation between the two superpowers took a back seat to what IGY boosters spun as healthy scientific competition between the communist East and the capitalist West.[22] The recent detonation of hydrogen bombs by both sides—the Americans' "Ivy Mike" in 1952, the Soviets' "Joe 4" in 1953—lent this competition a certain urgency.[23]

In the United States, popular concerns about nuclear weapons testing and the perceived gap between the quality of American and Soviet science gave American scientific leaders like Revelle, Berkner, and U.S. Weather Bureau chief Harry Wexler the latitude to push for a comprehensive program of geophysical research for the IGY—including research on CO_2 and climate. Funding for the American portion of the IGY came from Congress, primarily via the National Science Foundation. In May of 1957, members of the National Committee for the IGY—Wexler, Berkner, and Revelle among them—went before the House Appropriations Committee to ask for $39 million to run their eighteen-month program, including more than $2 million for oceanographic research and almost $3 million for research in meteorology.[24] In testimony that covered everything from the aridity of the Martian atmosphere to the depth of the Gulf Stream, Revelle introduced members of Congress to the possible relationship between an increase in CO_2 and an increase in global temperature, and he noted the many uncertainties that a program in CO_2 monitoring could hope to resolve. Congress approved the program, and ultimately IGY vessels and stations measured CO_2 at sixty locations around the world.[25] For eighteen months a full half century ago, CO_2 stood on the front lines of Cold War science.

FORECASTS AND MODELS

Revelle's growing interest in CO_2 paralleled other developments in climate-related atmospheric science that were also entangled in the web of Cold War research. These were developments in atmospheric modeling, in particular the advent of general circulation models.[26] Early numerical models of the earth's atmosphere grew out of efforts to use computers to forecast weather accurately during and after the Second World War, and in the ensuing decade scientists began to modify these models to attempt to predict changes in climate. Even more so than CO_2 monitoring efforts, models of the atmosphere had obvious applicability to questions about fallout and the potential distribution of radioactivity in the event of a nuclear exchange. In addition, some scientists and politicians linked modeling the weather and climate to controlling these geophysical forces, both for the domestic good via rainmaking and as a powerful Cold War weapon. Ultimately, the combination of the Scripps Institution's CO_2 research, the expansion of CO_2 monitoring during the IGY, and this government-sponsored effort to create a realistic, predictive model of atmospheric circulation all helped pave the way for the institutionalization of atmospherically oriented climate change research in America.

Climate science and meteorology both grew remarkably in the 1950s Cold War milieu, but the relationship between the two disciplines has not always been simple. "Weather," says the *Oxford English Dictionary*, refers to "the condition of the atmosphere (at a given place and time) with respect to heat or cold, quantity of sunshine, presence or absence of rain, hail, snow, thunder, fog, etc., violence or gentleness of the winds."[27] The characteristic weather conditions of a country or region over time, in turn, constitute that region's "climate."[28] As Paul Edwards explains in his history of computer-based climate modeling, *A Vast Machine*, however, until the 1950s the prime objectives and most important problems of climatology and meteorology were very different. Meteorologists have almost always sought primarily to provide a more accurate and reliable prediction of the weather, and in the early twentieth century, forecasting involved as much art as it did science. Weather forecasters—mostly amateurs—typically relied on local knowledge and local records to help them predict relatively small, short-term changes in local conditions for practical purposes. Climatologists, on the other hand, focused predominantly on reconstructing past climates and on using universal geophysical principles to try to explain long-term changes in those climates. Climatologists were, by and large, professional scientists working in university departments of geology or geography, and their work had few immediate practical applications.

This began to change in the 1950s, and it is helpful to approach the history of climate modeling with a brief history of meteorology in mind. As a scientific discipline, meteorology was relatively late to professionalize, but when this process finally began in the 1930s, the field and its methodology changed rapidly. Impetus for expanding American meteorology came from many quarters, including the American Meteorological Society, the Weather Bureau, the U.S. Navy, and the nascent American airline industry. The science, however, came primarily from Sweden. In the early twentieth century, a group of scientists studying under Vilhelm Bjerknes and his son, Jacob, in Bergen, Sweden, began to address weather phenomena in physical terms. They focused not just on traditional "weather types" and conditions on the ground but also on the movements of large masses of air and ocean that drove cyclones, storms, and, ultimately, just about every weather event under the sun. Their approach came to be known as dynamic meteorology.[29] In the late 1920s and 1930s, many of these "Bergen school" scientists began to appear as visiting scholars in the United States, and their work laid the foundation for the professionalization of American meteorology.

For academic meteorologists, the key to understanding the atmosphere lay in numerical weather prediction. The key to numerical weather prediction, in turn, lay in computers. In order to describe the dynamic processes of atmospheric motion, scientists had to solve long series of nonlinear equations, and they had to solve them quickly. During the First World War, British meteorologist Lewis Fry Richardson surmised that it would take about sixty-four thousand human computers solving equations continuously to keep up with the earth's weather as it unfolded—and that's to say nothing of predicting future weather.[30] During the Second World War, however, two electrical engineers working for the U.S. Army Ordnance Corps, Presper Eckert and John Mauchly, devised a machine that addressed this computing problem, the Electronic Numerical Integrator and Computer, or ENIAC.[31] Recognized as the first digital computers, ENIAC and its successor EDVAC (Electronic Discrete Variable Automatic Computer) were designed to compute the differential equations of ballistic weapons-firing tables more quickly, but they could also solve the differential and partial differential equations that described meteorological phenomena.

Following the suggestion of physicist Vladimir Zworykin (a pioneer in television tubes), meteorologists like Weather Bureau chief Francis Reichelderfer and his eventual successor Harry Wexler soon began to push for a computer-generated numerical weather forecast.[32] The result, after more than five years of research and development, was the interdisciplinary Joint Numerical Weather Prediction Unit. In March of 1950, this group used mathematician John von Neumann's new and improved computing machine, along with the expertise of European and American meteorologists, physicists, and mathematicians, to produce the world's first 12- and 24-hour computer-assisted numerical weather forecast.[33]

In the mid-1950s, as the IGY approached, scientists and their sponsors at the Weather Bureau and the National Science Foundation began to consider using models as a way to go beyond forecasting to a deeper understanding of the geophysical processes of the atmosphere. With 12- and 24-hour forecasts continuing to improve, meteorologists like Joseph Smagorinsky and Jule Charney at Princeton, as well as Victor Starr of MIT, turned their attention to general circulation models (GCMs), mathematical representations not just of the weather but of the movement of the atmosphere as a whole.[34] In 1955, Smagorinsky and Wexler launched a spin-off of the Joint Numerical Weather Prediction Unit focused specifically on GCMs. They called the unit the General Circulation Research Section.[35]

GCMs rely on two main elements to describe the global atmosphere. The first is called the dynamic core of the model, a framework of assumptions based on the physical properties of the atmosphere, as expressed through a set of nonlinear partial differential equations for fluid dynamic motion called the Navier-Stokes equations. The dynamic core of early GCMs typically revolved around the flow of energy, which GCM developers in the late 1950s had begun to explore through models of how the earth absorbed energy from the sun and reflected it back into space (energy budget models) and models of vertical energy transfers within and between the elements of the earth's atmosphere (radiative-convective models). Overlaid on top of this dynamic core is a second element of a GCM, the so-called model physics of all the other processes of the atmosphere, including the movements of air masses and atmospheric gases that affect weather and climate throughout the globe.[36] In the late 1950s, GCMs had already begun to provide very rough, low-resolution simulations of how the global atmosphere worked.

GCMs were not just about weather or climate, of course; like everything else in the early history of climate science, they were developed with Cold War applications in mind. As with CO_2 monitoring, general circulation modeling benefited greatly from the massive quantities of radioisotopes dumped into the atmosphere by nuclear weapons tests at known times and places throughout the 1950s. General circulation modelers traced radioactive CO_2 and other irradiated material through the atmosphere to provide observational data for their models.[37] Atmospheric models, in turn, provided a way to predict the distribution of radiation and fallout, as well as a method for detecting and even locating foreign nuclear testing activity.[38] The benefits of this kind of modeling for defense-related Cold War intelligence were not lost on anyone.

MASTERS OF INFINITY

Embedded in the government-sponsored efforts to model the atmosphere and forecast weather was an optimistic idea that if humans could better understand atmospheric processes, they could learn to control them. In 1946, General Electric's Nobel Prize–winning chemist Irving Langmuir and his colleagues Bernard Vonnegut and Vincent Schaefer pioneered small- and medium-scale weather-control projects that appealed to both military and civilian agencies. The discourse on weather and climate control quickly grew to involve both domestic and foreign policy interests and, again, in both realms Cold War security was never far below the surface.

Domestically, weather control projects promised to tackle real problems and, like most federal initiatives, to bring money to the states in which they were conducted. Boosters claimed that Langmuir's method of cloud seeding might disperse fog at airports, help alleviate droughts, or redirect dangerous storms—all of which appealed to members of Congress, especially those from the arid American West and the hurricane-prone Southeast. In the 1940s and 1950s, most of the applied weather-modification research in the United States was conducted with small-scale domestic projects in mind.[39] In 1953, Congress created an Advisory Committee on Weather Control to oversee these projects.[40]

The military—and especially the air force—was also interested in local and regional weather modification. Techniques that could benefit civilians at home might also be put to effective military uses abroad. Fog dispersal could help create an all-weather air force, for example. Rainmaking could slow an enemy's communications and supply routes—a strategy the air force later secretly employed during the Vietnam War.[41] More insidiously, the same techniques used to avoid drought could be used to create it, wreaking havoc on an enemy's food supply—a great asset in a nonshooting war.[42] The possibilities seemed endless, for as Langmuir took pains to point out, the energy available in cumulous clouds exceeded that of even an atomic bomb.[43]

In addition to these local and regional applications, larger-scale weather modification also had military and foreign policy implications, especially in terms of geophysical warfare more generally. In a paper presented to Congress in 1958, retired navy captain Howard T. Orville enumerated the ways in which humans might intentionally or unintentionally alter the weather or climate.[44] Coincidentally, one of his first points involved the unintentional warming of the earth through CO_2. Carbon dioxide was small potatoes compared to what might result from intentional weather modification, however. The Russians might spread colored pigments over the poles in order to absorb solar energy and ultimately melt the ice caps, causing a drastic change in the local and even global climate. They might dam the Bering Strait and use nuclear power plants to pump warm Pacific Ocean water into the Arctic Ocean, again melting the ice caps to provide the Soviet Union with warm-water ports at the expense of American coastal cities, which would be inundated. Finally, and perhaps worst of all, Orville cited German doctor Hermann Oberth, who foresaw "a gigantic mirror 'hung' in space" that would "focus the sun's rays as a giant magnifying glass at any desired intensity and beam" in order to protect orchards from frost, melt ice in ports, or serve any number of less innocent climatic purposes.[45]

Orville's paper—and the broader concern over weather and climate modification—reflected extreme anxiety over Russian scientific superiority in the wake of the 1957 *Sputnik* launch. Both government science boosters and congressional Democrats took advantage of this anxiety.[46] Lyndon Johnson in particular began to attack President Eisenhower for the administration's insufficient support of American science. Weather and climate control research was a prime area in which the United States must not get caught sleeping. An obscure Russian manuscript that appeared in English around 1961, *Man vs. Climate,* later revealed that by the late 1950s Russian scientists had in fact begun to conceptualize large-scale climate modification scenarios. Even so, hawkishness and hyperbole, rather than intelligence, pervaded the rhetoric of climate control in Cold War America.[47] "From space," declared Johnson at the Democratic Caucus in January of 1958, warning of the technologically and scientifically superior Russians, "the masters of infinity could have the power to control the earth's weather, to cause drought and flood, to change the tides and raise the level of the sea, to divert the Gulf Stream, and change temperate climates to frigid."[48]

Echoing his Democratic colleagues and the scientists who had their ears, Johnson advocated for more government funding, both for weather-modification research specifically and for science in general. Thus did weather and climate modification, alongside forecasting, general circulation modeling, and CO_2 monitoring, become central to the developing agenda of Cold War atmospheric science in America.

BIG SCIENCE, BROAD SCIENCE

As American atmospheric scientists began to articulate their field's research agenda in the late 1950s, they also began to think about what institutional structures they might need to pursue that agenda. In the 1950s, entrepreneurial scientists like Roger Revelle, Lloyd Berkner, and Harry Wexler capitalized on a spike in government interest in geophysical science to garner financial support for a wide range of projects in atmospheric science. As the decade came to a close, leaders in atmospheric science turned their attention to building a permanent home for these efforts. Their vision became the National Center for Atmospheric Research (NCAR, pronounced "n-car"), and, like the scientific pursuits it was meant to support, it bore the marks of its Cold War origins.

When you read "NCAR" on a page, it is easy to forget that the letters stand for a physical institution. In print, it blends with the era's myriad acronyms, a set of letters stamped on stationary indicating an elite atmospheric science club. But NCAR consists of real buildings that house real scientific and administrative equipment meant to serve real people. It also has a real geography, and thinking about the geography of NCAR is helpful for understanding how the institution reflects the peculiarities of atmospheric science as a Cold War pursuit in the 1950s and 1960s.

It may be more helpful to say that NCAR has *geographies,* or at least that the geography of NCAR is striking at a number of different levels. First and most noticeably, NCAR's headquarters, the Mesa Laboratory, is a geographical and architectural marvel in and of itself. The massive beige concrete structure sits high in the foothills of Boulder, Colorado, against the severe peaks of the Flatirons, blending remarkably well into the stark landscape. Designed by I. M. Pei, the building is a collection of vertical towers and labyrinthine corridors. The towers look imposing from the outside, but from the inside they are all but invisible unless you are actually in one of them and looking out its windows. Past the visitors' area at the main entrance, the Mesa Lab seems to have no center. Pei cultivated this sensation by design. The towers were meant to provide quiet spaces to work, but the rest of the building was made to foster casual interactions between scientists in communal spaces.[49] Pei consciously created a building without what he later called a "conventional scale" in order to foster an unconventional form of Cold War science.[50] The Mesa Lab did not open its doors until 1966, but its design and internal geography reflect the vision articulated by the institution's founders in 1960.

NCAR's geography extends beyond the architectural geography of the Mesa Lab. During my first visit to the institution in 2007, I went to the

Mesa Lab only once, to eat lunch and to see the lab itself. At that time, the NCAR archives were housed across town in the Foothill Lab, one of three other NCAR facilities in Boulder (there is a fifth NCAR lab at the Mauna Loa Observatory in Hawai'i and an NCAR supercomputing center ninety-two miles from Boulder in Cheyenne, Wyoming). Throughout the institution's history, the Mesa Lab has served as an impressive flagship, but NCAR as a whole has always remained at least partially decentralized, physically embedded in the city of Boulder, a network of offices and facilities connected by computers, telephones, and buses.

Less obvious than the internal geography of the Mesa Lab and the urban geography of NCAR labs in Boulder are two other geographies that speak to the institution's original objectives. The first is a human geography. NCAR is run by a parent institution, the University Corporation for Atmospheric Research (UCAR), and by charter it serves as a set of scientific facilities for professional atmospheric scientists working at colleges and universities throughout North America and at affiliated institutions abroad. The mix of permanent and visiting staff—and their numerous research priorities—reflects this mission. Today's NCAR houses a permanent scientific staff that rivals that of a midsized American university, but a revolving door of visitors—from technical staff to postdocs to senior scientists to the occasional historian—ensures that the institution maintains a scientifically cosmopolitan outward gaze. As those visitors return to their home institutions, they continually extend NCAR's scientific reach.

The other important NCAR geography is the geography of data that flow through the institution. NCAR is one of many nodes of data collection concerning earth processes, where literally millions of measurements find their way into computers running different models. Again, this is by design, following what the National Science Foundation (NSF) initially proposed for the institution in the late 1950s. On an informational level, NCAR operates in a global geography of data collection, a modern incarnation of a Cold War dream that NCAR itself was designed in part to fulfill.

Approaching NCAR through its institutional geographies provides a framework for understanding how the institution's structure reflects the context of its founding. But NCAR's modern geography alone tells only a part of this story. A return to the institution's history demonstrates more tangibly how the institution's geography reflects its Cold War roots.

In 1956, National Academy of Sciences president Detlev Bronk appointed a National Research Council Committee on Meteorology to "consider and recommend means by which to increase our understanding and control of the atmosphere."[51] The committee included Lloyd Berkner, Carl-Gustav Rossby, Jule Charney, John von Neumann, Edward Teller, Roger Revelle (in an informal advisory role), and a number of other luminaries of meteorology and Cold War science. The committee noted a shortage of training, personnel, and dedicated resources in the study of the atmosphere.[52] Meteorological research was fragmented and the discipline was underprofessionalized: 90 percent of American meteorologists were government employees trained by the air force during the Second World War, and the vast majority of these more than twenty thousand military men had no doctorates.[53] The nature of the subject matter, meanwhile, required a sophisticated interdisciplinary approach, often using expensive, large-scale equipment that was at the time unavailable to most scholars.[54] In early 1958, with the International Geophysical Year underway, the committee recommended that the federal government increase overall financial support for basic atmospheric research and that the NSF underwrite the establishment of a National Institute of Atmospheric Research (NIAR, later changed to NCAR) to coordinate and conduct this research.[55]

Though the Committee on Meteorology recommended federal funding for NIAR, it strongly suggested that the new institution be run by representatives from meteorology departments at American universities rather than by government bureaucrats. The committee's vision for a university-based research program reflected an ambivalence about the broader U.S. Cold War research system. On the one hand, the committee saw NIAR as a way to take advantage of government money in order to address a constellation of scientific, civilian, and military concerns, all at least ostensibly in the service of the state. Committee members recognized that only the federal government could provide the resources and facilities scientists needed to approach the "fundamental problems of the atmosphere on a scale commensurate with their global nature and importance."[56] Airplanes loaded with atmospheric-monitoring devices; expensive ground-based instruments, including powerful telescopes and coronographs; and access to new technologies like computers and satellites outstripped the financial capabilities of even the most well-endowed universities. Funding would have to come from the government research establishment—the Department of

Defense, Office of Naval Research, and National Science Foundation—and NIAR in that sense would itself be a government institution.

On the other hand, atmospheric scientists feared the hierarchy, secrecy, disciplinary compartmentalization, and strict oversight that might accompany government support. In particular, they worried that the military would come to dominate atmospheric research.[57] By channeling federal resources into an institution expressly created to serve the interests of universities, atmospheric scientists could insulate themselves from the direct oversight of their government sponsors. In February of 1958, the Committee on Meteorology began working closely with the University Corporation for Atmospheric Research—leaders from the nation's university meteorology departments—to establish the structure and goals of NIAR.[58] In early 1959, UCAR officially became the parent organization of the new institution, by then renamed NCAR. The UCAR board of trustees submitted a detailed proposal to the NSF that outlined NCAR's purpose, institutional structure, and initial research plans—a proposal published as NCAR's founding document, the "Blue Book." The Blue Book also contained the original sketches of what later became the physical geography of I. M. Pei's Mesa Lab.

The Blue Book described a unique variation on the "big science" that characterized government-sponsored research in other fields—especially high-energy and particle physics—during the Cold War.[59] Coined in 1961 by Alvin Weinberg of the Oak Ridge National Laboratory, the term *big science* typically describes expensive, large-scale, government-sponsored research carried out either by private corporations or large scientific institutions. UCAR certainly wanted NCAR to operate at a large scale; in fact, the global nature of the subject matter all but demanded thinking big, and the Blue Book explicitly stated that NCAR should provide "facilities and technological assistance beyond those that can properly be made available at individual universities."[60] As Bruce Hevly notes in "Reflections on Big Science and Big History," however, big science in the twentieth century was typically organized hierarchically, with vertically integrated teams striving for institutionally directed scientific objectives.[61] This was not the case at NCAR. If a typical big-science project—the Berkeley cyclotron, for example—operated like a corporation, where a CEO or board of directors dictated research priorities with specific goals in mind and then hired scientists to do the job, NCAR operated more like a cooperative, where visiting scientists from member universities could have access to and control over facilities otherwise unavailable to them. As an administrative necessity, the Blue Book divided NCAR into three main sections along disciplinary lines—physical, mathematical, and chemical laboratories—but these divisions and the hierarchies within them were intended to be flexible. A genial, well-liked, hands-off director, Walter Orr Roberts, nurtured that institutional flexibility. The semipermanent nature of NCAR's scientific staff, meanwhile, composed largely of self-directed visiting scientists and postdoctoral fellows, further subverted the institution's administrative and scientific hierarchies.[62]

The scale and complexity of the atmosphere itself added to the singularity of the "big" atmospheric science envisioned for NCAR. Most big-science projects in America revolved around an object or a single, tangible objective—the cyclotron, the atomic bomb, the gravity probe, or the space telescope. These projects' managers typically approached complex systems with relatively single-minded goals—to collide particles or to collect astronomical data—and they tended to treat their projects and experiments in isolation from the outside world. By contrast, atmospheric scientists launched NCAR as a way to study the composition, motion, and behavior of the various elements of a global space that in fact *was* the outside world. UCAR's board of trustees presented NCAR as a sort of base station in a much larger effort to study the atmosphere at its full, global scale.[63] Simply describing the atmosphere—let alone predicting or controlling it, as atmospheric scientists hoped to do—required not only the data-collection facilities and instruments available to NCAR locally but also access to monitoring stations, and later satellites, around the globe. Atmospheric scientists used NCAR's computers to generate radiation budget models and global circulation models, for example, but they could do so only using data generated elsewhere (mostly during the International Geophysical Year). NCAR was intended as "an international intellectual center with a marked interdisciplinary flavor" capable of serving as a focal point in a developing field of science that was not only "big" in personnel and resources but also geographically and disciplinarily broad.

CO₂ AND CLIMATE AT NCAR

The opening of the NCAR labs in Boulder in 1960 represented an institutional confluence of four overlapping strands of atmospherically oriented climate research: radiation budget modeling, general circulation modeling, the study of weather and climate control, and the CO_2 question. Each avenue of inquiry benefited from the peculiar human and technological resources of the new institution, and concern over atmospheric CO_2 and climate change at NCAR grew with the institution itself.

Work on climate at NCAR began with modeling. Scientists continued to search for a more complete model of the atmosphere, and that model began with energy—or, more precisely, with energy budget models (mathematical descriptions of the flow of energy from the sun, typically including the distribution and flux of particle energy, short- and long-wave radiation, and heat exchanges between the atmosphere and the earth's surface through an otherwise static atmosphere). As atmospheric scientists knew, these radiation models severely complicated existing models of atmospheric circulation.[64] To effectively incorporate radiation equations into general circulation models, atmospheric scientists needed more information on worldwide sources and sinks of solar energy. Fortunately, as the Blue Book pointed out, "the imminent advent of meteorological satellites will place in the hands of the atmospheric scientist a new and remarkably powerful tool for viewing the heat balance problem in its global entirety."[65] Another new Cold War institution, the National Aeronautics and Space Administration (NASA, created in 1958), was already hard at work on these satellites.[66] NCAR's scientists sought to reap the benefits of the new data.

This was not purely science for its own sake. To atmospheric scientists and their government sponsors, fundamental knowledge of the earth's energy budget was the keystone of efforts to control atmospheric processes; and as Lyndon Johnson's support for weather-modification research underscored, controlling atmospheric processes appealed to a broad domestic constituency. The Blue Book stated the case plainly: "The physical linkage between the heat budget and the general circulation of the atmosphere is such a close one that any hope of effective climate control is likely to lie in alteration of some aspect of the heat budget."[67]

An increase in CO_2 was one thing that scientists thought might alter the heat budget. If the Blue Book expressed atmospheric scientists' optimism about the possibilities of eventually controlling the climate, it also revealed the seeds of concern about unintentional climate modification, and CO_2 stood out as a potential cause for concern. Atmospheric scientists identified atmospheric CO_2 as one of several poorly understood physical and chemical phenomena that could complicate their models and understanding of the global atmosphere. As they did elsewhere, in the Blue Book these scientific leaders associated questions about CO_2 with questions about nuclear testing: "Man's activities in consuming fossil fuels during the past hundred years, and in detonating nuclear weapons during the past decade have been on a scale sufficient to make it worthwhile to examine the effects these activities have had upon the atmosphere. Reference is made here to the still unsolved question of whether the carbon dioxide content of the atmosphere is increasing as a result of combustion processes and the even more elusive question as to possible changes in the earth's electrical field as a result of nuclear explosions."[68]

The plan to study CO_2 and climate at NCAR was hardly a clarion call warning of global warming. It was, like Roger Revelle's famous "geophysical experiment" comment, primarily a statement of scientific uncertainty. Even so, by 1963, it was difficult for scientists to look at the Keeling Curve without scratching their heads. If NCAR's structure reflected atmospheric scientists' doubts about certain aspects of the Cold War research system, when atmospheric scientists made CO_2 a programmatic focus at the new institution they also sowed their specific doubts about CO_2 into the field's institutional soil. Both at NCAR and elsewhere, in the 1960s these seeds of doubt about atmospheric CO_2 began to germinate. They were watered, in part, by Charles David Keeling himself.

THE "INVISIBLE POLLUTANT"

Just as it was Keeling's bosses, rather than Keeling himself, who put CO_2 on the Cold War research map, so too did Keeling's scientific superiors articulate a connection between CO_2 and pollution before Keeling put the problem succinctly in 1963. As early as 1957, when Roger Revelle testified before Congress on CO_2 monitoring, the handful of atmospheric scientists interested in CO_2 monitoring had used the potentially disruptive impacts of CO_2 on climate to lobby for financial support. A rise in global CO_2 could raise temperatures, they argued, which in turn could melt glaciers, leading to a rise in sea level. Changes in temperature might also lead to changes in weather that would affect agricultural output in certain regions—for the better in some places, for the worse in others.[69] Making his case before western members of Congress in 1957, for example, Revelle suggested that minor changes in the earth's climate could threaten the water supply of the American Southwest.[70] Revelle and other scientists emphasized that these were mere conjectures—vague possibilities whose likelihood remained unknown but whose potential consequences made the curious question of CO_2 accumulation worth studying. "Only God knows whether what I am saying is true," Revelle admitted to his congressional audience. "What I am driving at is that this business of the carbon dioxide production is in fact a way of studying climatic changes."[71]

In the late 1950s and early 1960s, scientists began to chip away at some of the basic uncertainties of atmospheric CO_2. In 1958, Swedish meteorologists Bert Bolin and Erik Eriksson helped clarify the nature of the carbon dioxide exchange between the oceans and the atmosphere that Revelle and Suess had discussed briefly in their *Tellus* article the previous year. Bolin and Eriksson used a short-term model of atmosphere-ocean interactions that included a more sophisticated account of the buffering mechanism referred to by Revelle and Suess. The Swedish researchers predicted that atmospheric CO_2 would most likely increase by 25 percent by the end of the century.[72] A recalibrated measurement of the Suess Effect, alongside Keeling's steadily rising CO_2 curve from his measurements at the Mauna Loa Observatory, seemed to confirm that atmospheric CO_2, as Guy Stewart Callendar had suggested years before, was in fact rising.[73] It also appeared that humans were the primary cause.

Developments in CO_2 research overlapped with the broader scientific community's increased awareness of and concern for related environmental issues. In the late 1950s and early 1960s, scientists from a variety of fields began to use their expertise to study the human and environmental impacts of the great scientific and technological achievements of the twentieth century. Most famously, in 1962 marine biologist Rachel Carson published *Silent Spring*, an exposé of the environmental and public health impacts of chemical pesticides and one of the fundamental texts of modern environmentalism.[74]

Few scientists took the same approach or had the same impact as Carson. Roger Revelle's career path was more characteristic of most geophysical scientists' involvement in environmental issues.[75] His chief environmental concerns involved nuclear radiation, population, and pollution. Building on his earlier work with the navy, in the mid-1950s Revelle chaired the National Academy of Sciences Committee on the Biological

Effects of Atomic Radiation, which released a report in 1956 (the so-called BEAR report).[76] In 1961, he left the Scripps Institution to work as science advisor to Secretary of the Interior Stewart Udall, perhaps the most vocal supporter of conservation in the Kennedy and Johnson administrations, responsible for helping environmentalists with several initiatives: the Great Swamp National Wildlife Refuge in New Jersey, the Clean Air Act of 1963, the Wilderness Act of 1964, and the Endangered Species Act of 1966. In 1964, Revelle moved to Cambridge to head Harvard's new Center for Population Studies, where he focused on the "consequences of population change on human lives and societies, and of the biological, cultural, and economic forces that influence human fertility."[77] Revelle was not necessarily an environmentalist; rather, he shared with his colleagues a commitment to the sensible scientific management of the environment and natural resources for the benefit of humankind.

A number of atmospheric scientists echoed Revelle's personal and professional interest in environmental issues, and in the early 1960s they began to address CO_2 as it related to pollution, a buzzword of the environmental movement. Initially, Revelle and his colleagues drew a loose association between CO_2 research and air pollution, much the same as they had drawn an association between CO_2 and radiation since the mid-1950s. For example, in the 1962 report *The Atmospheric Sciences, 1961–1971*, commissioned by President Kennedy's science advisor Jerome Wiesner, the National Academy of Sciences' Committee on Atmospheric Sciences proposed to study water vapor and CO_2 alongside other trace gases—ozone, methane, oxides of nitrogen and sulfur, and "several radioactive gases"—associated with industrial air pollution.[78] The relationships between these gases, as well as their potential impacts, remained vague, but the report implied that CO_2 accumulation, inadvertent weather modification, atmospheric radiation, and air pollution were all of a piece.[79]

As a pollutant, however, CO_2 was problematic. CO_2 was a good indicator of smog, haze, and other forms of pollution associated with fossil fuel consumption, but as Charles David Keeling noted in a piece he wrote for a Conservation Foundation conference in 1963, "air pollution in the ordinary sense does not include the CO_2 rise in the atmosphere."[80] Whereas substances like DDT, strontium 90, or sulfur dioxide directly threatened the health and well-being of ecosystems and human populations, CO_2 was a natural product of animal respiration and a gas essential to photosynthesis—and, by extension, essential to life on earth. Both the Air Pollution Control Act of 1955 and the Clean Air Act of 1963 dealt with toxic or noxious substances that had direct, typically local effects on human health.[81] These substances emerged primarily from discernable point sources—cars or factories—that government officials could regulate. CO_2, by contrast, was harmless at the local level. Only through complex geophysical processes resulting from a global accumulation of the gas could CO_2 impact human populations—hardly a phenomenon that state or federal governments were in a position to deal with.[82]

Keeling's take on CO_2 came a long way between 1955 and 1963. In 1955, Keeling was a newly minted chemistry Ph.D. walking in the woods around Big Sur, probably trying to avoid his Cal Tech postdoctoral advisor Harrison Brown and thinking about how to translate measuring ground-level CO_2 and its isotopes into camping trips in the forests of California. Keeling's interest in CO_2 revolved around daily cycles of local plant respiration, and at twenty-three years old he was beginning to be known as an expert on the chemistry of forest air.[83] By 1963, Keeling was thinking much bigger, reinterpreting CO_2 at a global level in a new way and for a new audience.

Still interested in forests and still passionate about spending time outdoors, the Keeling of 1963 saw important physical and philosophical connections between the increasingly popular concern over pollution and the scientific interest in CO_2 accumulation. Increased atmospheric CO_2, Keeling argued, might ultimately act similarly to pollution, albeit through a more complex chain of causation. CO_2 could threaten human welfare through a rise in sea level and might cause significant ecological and agricultural disruption by warming the air and ocean sufficiently to change patterns of plant growth and species distribution.[84] CO_2 interacted with other fossil fuel pollutants—particularly sulfur dioxide—in unpredictable ways, further increasing the unpredictability of humans' impact on the atmosphere.[85] "The potentially dangerous increase of CO_2, due to the burning of fossil fuels, is only one example of the failure to consider the consequences of industrialization and economic development," Keeling wrote.[86] "It is almost impossible to predict all of the consequences of man's activities. It is possible, however, to predict that there will be problems."[87]

In 1963, Keeling described CO_2 as a type of pollution, framing it within the context of environmental problems for the first time. In 1965, the President's Science Advisory Committee (PSAC) attempted to catalogue these problems—and potential solutions to them—in a comprehensive executive report, *Restoring the Quality of Our Environment*.[88] Commissioned by Donald Hornig, President Johnson's special assistant for science and technology, the PSAC Environmental Pollution Panel broadly defined pollution to encompass a loose constellation of general and specific issues, and CO_2 made the cut. The report tackled the public health impacts of air pollution, water pollution, and pesticides; the impact of human actions on other living organisms; the "impairment of water and soil resources"; the "polluting effects of household detergents"; urban decay; and the climatic effects of CO_2.[89] "Environmental pollution," the report read, "is the unfavorable alteration of our surroundings, wholly or largely as a by-product of man's actions, through direct or indirect effects of changes in energy patterns, radiation levels, chemical and physical constitution and abundances of organisms. These changes may affect man directly, or through his supplies of water and of agricultural and other biological products, his physical objects or possessions, or his opportunities for recreation and appreciation of nature."[90] According to the PSAC authors, though it affected humans' well-being only indirectly through the mechanisms of climatic change, CO_2—the "invisible pollutant"—nevertheless constituted a potentially dangerous by-product of advanced industrial society that needed to be monitored and potentially counteracted in the future.[91]

Despite their growing concerns about the accumulation of CO_2 from fossil fuels, however, geophysical scientists in the 1960s remained characteristically optimistic about their ability to counteract the gas's climatic impacts through intentional geophysical modification. Roger Revelle served as the chair of PSAC's carbon dioxide group, and as usual he pointed out the pressing need for better understanding of atmospheric phenomena—understanding that could lead to control.[92] The first step was to build a more comprehensive model of the atmosphere, a model that Revelle predicted the Weather Bureau would have up and running in less than two years.[93] The understanding gleaned from such models would ultimately allow humans to make conscious changes to other processes affecting climate. Revelle was particularly sanguine about plans to change the earth's albedo—the sun's energy reflected back into space by the surface of the earth—by scattering reflective particulates over swaths of the ocean.[94] If Keeling's 1963 article postulated CO_2 as an environmental problem for the first time, Revelle's response to that reinterpretation of CO_2 was an early articulation of the most important refrain in the political history of global warming: more and better science, applied in the right way by the right people, would solve the problem.

A CHILD OF THE COLD WAR

It would be difficult to overstate the impact of the Cold War on the development of atmospheric science and CO_2 research in the United States. The juxtaposition between the experiences of Guy Stewart Callendar, the obscure British steam engineer, and those of Charles David Keeling and Roger Revelle underscore the point. In 1938, Callendar contended that human consumption of fossil fuels was, via the accumulation of CO_2 in the atmosphere, slowly raising the mean temperature of the earth. Fossil fuel combustion, he showed, had added about 150 billion tons of CO_2 to the atmosphere, roughly two-thirds of which remained aloft. Using the known radiation absorption coefficients for water and CO_2, he estimated that this increase in atmospheric CO_2 should cause the world to warm at a rate of about 0.003°C per year—a figure not far from the 0.005°C rate of warming measured by meteorologists in the half century before Callendar's paper.[95] That is to say, (1) CO_2 was rising; (2) increased CO_2 would lead to warming; and (3) the earth was in fact getting warmer.[96] "Few of those familiar with the natural heat exchanges of the atmosphere," Callendar wrote, "would be prepared to admit that the activities of man could have any influence upon phenomena of so vast a scale."[97] Nevertheless, he contended, "such influence is not only possible, but is actually occurring at the present time."[98] The earth, he declared, was warming, and humans were responsible.

In retrospect, Callendar's contention about anthropogenic warming seems prescient. Callendar was right. And yet in his own time, Callendar's claim garnered little attention outside of a small community of meteorologists; and among this small group, few took the notion of human-caused global change seriously.

By 1965, however, atmospheric scientists had incorporated CO_2 research into individual projects and institutional directives at the national and international levels, and CO_2 had appeared at least twice on the agenda of the president of the United States.[99] In part, the increased visibility of CO_2 and climate arose out of advances in science itself. Papers in the 1950s and

1960s on the physics and chemistry of CO_2 by Gilbert Plass, Roger Revelle and Hans Suess, and Bert Bolin and Erik Eriksson, for example, alongside better models of the general circulation and the earth's radiation budget developed by Jule Charney, Joseph Smagorinsky, and others began to paint a clearer picture of the relationship between CO_2 and climate.

These scientific developments dovetailed with philosophical, financial, and structural changes in atmospheric science that were ultimately just as important as science itself in putting CO_2 on the popular and political map. These changes were by and large products of the Cold War, and they begin, as many Cold War stories do, with the bomb. The explosion of the first atomic bomb in Alamogordo, New Mexico, in 1945 established humans as agents of large-scale geophysical change in a new way, and it opened the door for the possibility that human actions might already be altering geophysical processes unintentionally. Callendar suggested that his audience would not be ready to admit that humans could affect the earth on such a large scale. After Alamogordo, Hiroshima, and Nagasaki—to say nothing of the H-bomb tests of the mid-1950s—far fewer people would be so quick to deny humanity's geophysical agency.

Perhaps more important, the advent of the bomb fostered the development of an expansive science infrastructure, with the financial and technological resources necessary for studying the earth's large-scale geophysical systems at a scale beyond what amateur scientists like Callendar ever could have imagined. Leaders in atmospheric and other geophysical sciences like Roger Revelle, Harry Wexler, Lloyd Berkner, and Jule Charney soon began to capitalize on newly available scientific resources by framing research on CO_2 and climate in terms of its potential geopolitical significance. These scientists promoted atmospheric science and CO_2 research through individual projects like Keeling's Mauna Loa monitoring stations and the global monitoring effort of the International Geophysical Year and through the founding of institutions of atmospheric science, most notably NCAR. These scientists' successes in incorporating CO_2 research, atmospheric modeling, and studies of weather and climate control into the agenda of government-sponsored Cold War science fueled a growing interest in and concern over CO_2 in the early 1960s.

Many atmospheric scientists hoped to tackle CO_2 and atmospheric circulation on a worldwide basis, but interest in atmospheric science was by no means exclusively global. The short-term efforts to predict and control weather and climate appealed primarily to politicians from the arid West and hurricane-ridden Southeast who recognized the tangible benefits of understanding meteorological events. Still, even domestic concerns about weather and climate entwined with the scientific objectives of the Cold War. Scientific leaders interested in the global phenomena of the atmosphere used the potential regional impacts of these phenomena to gain support from western and southeastern senators for their research. Senators from these regions in turn cited a patriotic concern for the national security implications of weather and climate control—alongside the benefits of local weather-modification projects—in order to win appropriations for further research. Boosters of atmospheric science emphasized these potential applications of basic research, especially when it came to the nebulous issue of CO_2 accumulation. The combination of politicians' interest in regionally specific weather-modification projects and the federal government's broader interest in the potential geopolitical impacts of atmospheric change helped men like Revelle, Berkner, and Wexler establish atmospheric science in general—and research on CO_2 and climate in particular—as a Cold War research priority for the 1960s.

At the outset of the decade, atmospheric scientists' vision for their field was justifiably optimistic. Both their scientific and administrative success, largely made possible by newly available government money, fed this optimism. Relatively rapid and continuous improvements in weather forecasting and global circulation models, along with better and more complete theoretical descriptions of atmospheric motion, convinced many atmospheric scientists—especially meteorologists and atmospheric modelers—that the primary obstacles to predicting and eventually controlling the weather and climate were the availability of accurate global data and computers' capacity to handle that data. New Cold War technologies—especially the satellite and the increasingly powerful computers developed in the late 1950s and early 1960s—promised to remove both of these potential roadblocks. Beginning with NCAR, leaders in atmospheric science sought to create more permanent institutional arrangements for scientists to access these new tools.

Led by Jule Charney, American atmospheric scientists in the early 1960s also sought to expand the geographical reach of their research by institutionalizing the system of global atmospheric monitoring begun during the International Geophysical Year.[100] With the International Council of Scientific Unions and the World Meteorological Organization (both connected to the United Nations), Charney began work on the Global Atmospheric Research Programme (GARP), a comprehensive system of monitoring and data sharing endorsed by President Kennedy that finally went live at the end of the decade.[101] Scientists continued to emphasize the uncertainties and outstanding questions of their research, but they saw answers on the horizon.

What they did not see, however, was that for the next half century each round of scientific answers would yield increasingly complex scientific and political questions—questions that would force them to rethink the meanings of CO_2 in increasingly complicated contexts. In 1963, some of these new questions were already stirring.

To begin with, atmospheric scientists' optimism about the future of their profession masked an undercurrent of anxiety about the Cold War research system that made their success possible. Like many of their colleagues in other disciplines, they worried about the centralization and militarization of scientific research in the service of the state.[102] The character and administration of their institutions—particularly NCAR, but also later GARP and, to an extent, the Geophysical Fluid Dynamics Laboratory—reflected their efforts to insulate themselves from the very government agencies that provided financial support for their research. As a result, atmospheric science became a particular form of "big science," unique not only in its global reach but also in its flattened management hierarchies and user-driven research agenda.

Heightened concerns over the accumulation of atmospheric CO_2 in the early 1960s also reflected atmospheric scientists' growing ambivalence about the social and environmental impacts of science itself. Philosophically, scientists like Roger Revelle and Charles David Keeling shared a general interest in the unintended consequences of human actions with America's bourgeoning environmental movement, and the two communities' concerns had common Cold War roots. In the early 1960s, atmospheric scientists even began to frame CO_2 in terms of the dominant paradigm in contemporary environmental politics: the paradigm of pollution. But atmospheric CO_2 was an unconventional pollutant, and the scientists who studied it maintained a faith in their discipline's ability to study, understand, and ultimately solve environmental problems that many of their environmentalist counterparts had begun to doubt. Nevertheless, the precautionary ethos that led atmospheric scientists to pursue CO_2 as an environmental pollutant put them in loose conversation with the growing American environmental movement in the 1960s. Atmospheric scientists were hardly environmental activists in the mid-1960s, but by the end of the decade the scientific, political, and epistemological pieces were in place for that to change. Already, a new interpretation of the Keeling Curve was at hand.

2

SCIENTISTS, ENVIRONMENTALISTS, AND THE GLOBAL ATMOSPHERE

IF YOU LOOK CLOSELY AT THE DATA BEHIND THE KEELING CURVE, YOU will notice that the undulating line representing the annual cycles of atmospheric CO_2 over time is not continuous. In the spring of 1964, there is a gap. In January of 1964, near the midpoint of the oscillation that peaks in May and bottoms out in October, global atmospheric CO_2 stood at roughly 319 ppm. In May, at the peak, it was 322 ppm. And between those two points, nothing. Charles David Keeling's funding had run out, and for three months in early 1964 there was insufficient data to provide a monthly average of atmospheric CO_2 at the Mauna Loa Observatory. In May, new measurements began.

There is a similar divide between old and new in the history of climate change in the 1960s, though it is perhaps less obvious and certainly less abrupt. Before the mid-1960s, atmospheric CO_2 remained a scientific curiosity whose potential impacts on global temperature and sea level enabled science advocates like Keeling and Roger Revelle to secure research funding. Their concerns were speculative, balanced by their optimism that scientific and technological developments could help solve whatever human or environmental problems CO_2 might present. By the end of the decade, however, scientists and the general public had begun to see atmospheric change—including rising CO_2—in a new way. The optimism about the "grand experiment" of CO_2 faded in the 1960s, and some scientists became increasingly concerned about the environmental impacts of atmospheric change and dubious of the beneficence of Cold War science and technology. The confluence of these worries forced scientists to rethink the meaning of atmospheric CO_2. The vehicle for this reinterpretation was an airplane: the supersonic transport.

On June 5, 1963, the Kennedy administration announced a large-scale cooperative program between industry and government to build a commercial passenger aircraft that would travel faster than the speed of sound. As a bold statement of America's commitment to superiority in aerospace technology during the Cold War, the American supersonic transport, or SST, initially found wide support in Congress and among the public. Over the course of the 1960s, however, Kennedy-era military-industrial aerospace projects fell out of favor with an increasingly skeptical public, and support for the SST waned.[1] Critics questioned the wisdom of the expensive program in light of increased spending on the Vietnam War and a slowdown in the growth of federal science budgets. Environmentalists, growing in influence in American politics in the 1960s, argued that the marginal benefits of the plane could not justify its fuel consumption, its noise, or the disruptive sonic boom it left in its wake.

Environmentalists' concerns revolved around the threats SSTs posed to Americans' everyday quality of life—noise pollution and sonic booms—but as the debate over the plane progressed, scientists also began to question its possible impact on the global environment. Researchers began to study the long-term, large-scale effects of SST operation on the earth's atmosphere, and soon atmospheric change—including rising atmospheric CO_2—emerged as a meaningful new type of global environmental issue. As it did, atmospheric scientists also emerged as a new voice in environmental politics. The peculiarities of their science-first environmental advocacy would develop into one of the most important themes in the political history of global warming. The SST controversy thus laid the groundwork for a new way of understanding CO_2 in the environmentally conscious historical context of the 1970s.

THE MULTIPLE DEATHS OF THE SST

To understand the distinctions between mainstream environmental activism and the science-first form of climate advocacy developed in the

1960s and 1970s, it is helpful to tell the story of the death of the SST twice. The failure of the SST program unfolded in two parts—once in 1968 and once in 1971—and the debate, like the larger story of CO_2, went through multiple iterations. The program's first defeat occurred at the hands of a noise-abatement bill in 1968, and the second, more resounding budgetary defeat in 1971 was in one sense a continuation of this first, familiar story of successful environmental activism. However, a retelling of the 1971 budgetary defeat with a new component in mind—atmospheric change—reveals another dimension of the narrative. It underscores how the SST introduced atmospheric change as a new kind of environmental problem tied to a peculiar form of environmental advocacy undertaken by an odd new breed of environmentalist.

The first death of the SST began early in 1967, when engineers working on Boeing's Model 733-390, the SST aircraft recently approved by the Federal Aviation Administration (FAA), ran into a snag. The "swing-wing" mechanism that allowed for variable wing positions—a key feature of the new plane's design—was too heavy. The extra weight would reduce the SST's range and thus compromise its utility as a transcontinental aircraft, the niche its proponents had carved out for it in the larger aviation market. What's more, the wings threatened to overflex the extralong fuselage (the longest ever built), compromising control. The wing design, Boeing finally decided in 1968, would have to be scrapped, the project delayed. Unbeknownst to Boeing, that delay would in the end help kill the American supersonic program altogether.

The SST had its critics before Boeing encountered this problem, but until 1967 the plane's status as a Kennedy-era aerospace project helped insulate the SST from both congressional and media opposition.[2] It was, in the words of the London Observer's aviation correspondent Andrew Wilson, a "political aeroplane," and it carried with it a set of assumptions about the role of technological development in foreign policy and domestic economic stability.[3] In the post-Sputnik aerospace technology boom of the early 1960s, few senators had incentive to block a program its promoters claimed would create American jobs, bolster the American economy, and ensure American hegemony in aircraft manufacturing in the face of international competition. Washington State, home to the SST's primary design firm, Boeing, and a major hub of U.S. aircraft manufacturing, stood to benefit significantly from the project. The state's powerful Democratic senators, Warren Magnuson and Henry "Scoop" Jackson, lent their strong support. Meanwhile, the Anglo-French Concorde and the Russian TU-144 seemed likely to hit production by the early 1970s. Proponents of the SST reiterated time and again that if Americans did not build supersonic aircrafts, then they would wind up buying them.[4]

In addition to Boeing's design problems, however, the SST had serious nontechnical issues, and in the delay caused by the wing flaw these issues came to light. First and foremost, the economics of the SST were at best questionable. As Swedish aviation consultant Bo Lundberg wrote in a 1965 *Bulletin of the Atomic Scientists* article (and later testified to Congress), while the SST might create a few jobs for highly skilled workers already in high demand, the project held little promise of bolstering the U.S. economy.[5] The SST might not even break even. Lundberg predicted that the Concorde, the model for the American SST, at the projected cost of $46 million per plane and between three and four times the fuel consumption per seat mile of a standard subsonic jet, would serve an "appallingly small" market. In the end, he predicted, the Concorde would lose money for its British and French sponsors. (It did.)[6] As Defense Secretary Robert McNamara confirmed in an internal White House study shortly after Lundberg's article came out, the American SST might fare even worse.[7] By 1967, buying the French model instead of building an American one for a loss did not look so bad.

Perhaps more important, as opposition to the SST mounted, the fight against the plane began to represent a larger challenge to the mind-set that had made the SST a possibility in the first place.[8] Though perhaps not as obvious or abrupt as the gap in the Keeling Curve, at some point in the mid-1960s a rift began to develop between the technological optimism of the Kennedy years and the skepticism characteristic of the tumultuous final years of the decade. By the time the SST was defeated in 1971, Democratic senators cited the plane as a "symbolic issue in the struggle over new priorities and directions for the nation."[9] Much of the impetus for this struggle for new directions came from profound doubts about the integrity of government and the role of an overly powerful technology sector tied directly to America's controversial military engagements. In the case of the SST, opposition that began as an economic argument against irresponsible aerospace spending came to fruition as growing concern for the quality of America's natural and human environments.

For many environmentalists, the supersonic transport represented the kind of familiar backyard threat to Americans' everyday quality of life that Sam Hays later identified as the driving concern behind the postwar American environmental movement in his touchstone 1987 *Beauty, Health, and Permanence*. Among its many liabilities—the SST was louder, less fuel efficient, and more costly than any other aircraft of its day—citizens' groups and environmentalists worried most about the deafening shock wave the aircraft would leave in its wake: the sonic boom. Traveling at speeds exceeding Mach 1 (the speed of sound, about 340 meters per second, or 760 miles per hour at sea level), the SST promised to leave a rumbling blanket of disruption over the landscape below, a sharp and crackling clap that would shake windows and in some cases even damage buildings.[10] As a bicoastal commuter jet, the SST would visit this disturbance upon thousands of communities across the nation on a daily basis.

In 1967, popular opposition to the SST coalesced around two Harvard professors, biologist John Edsall and physicist William Shurcliff. Influenced by Lundberg, who by 1966 had expanded his criticism of the SST to include the sonic boom issue, Shurcliff founded the Citizens League against the Sonic Boom, with himself as director and Edsall as his deputy.[11] Shurcliff lobbied Democratic members of Congress to pressure the FAA and NASA to reassess the supersonic transport in light of tests conducted over Oklahoma City that demonstrated just how disruptive daily sonic booms could be.[12] Shurcliff complained that the tests had not been adequately considered in NASA's pro-SST report on the sonic boom, that in fact the tests' results undermined the cost-benefit analysis behind NASA's support.[13]

While Shurcliff lobbied Congress to reevaluate the Oklahoma City tests, he also began to pitch the sonic boom problem to American environmentalists—particularly the Sierra Club and the Wilderness Society. In order to sell the cause to organizations like the Sierra Club and its affiliated environmental organizations, Shurcliff noted that sonic booms would disturb wildlife and the tranquility of wildlands. Secretary of the Interior Stewart Udall echoed Shurcliff's concerns on the front page of the *Washington Post*, where he publicly "decried the boom's impact on wildlife, national parks, and adobe structures built in the Southwest by Indians."[14] Though marginal compared to more prominent environmental affronts like logging and mining, the threat that sonic booms presented to both Indian heritage and the American wilderness experience resonated among many members of U.S. environmental groups.

The Citizens League's fight against the SST played out within the familiar patterns of legislative lobbying, litigation, and environmental monitoring that have come to characterize modern environmental politics. Like the bulk of the environmental activism of the period, opposition to the sonic boom began with a grassroots organization representing a large constituency of concerned citizens and flowered into a national campaign to protect Americans' quality of life. The Sierra Club, the Wilderness Society, and later Friends of the Earth, Environment!, the Natural Resources Defense Council, and other major environmental groups all moved to oppose the SST. Their campaigns achieved notable success.

Environmentalists also quickly expanded their criticism of the SST, turning from sonic booms to the related issue of airport noise pollution. In addition to producing sonic booms in flight, the SST engines roared at decibels well above the threshold dictated by noise regulations on the ground. Citizens' groups had already persuaded the FAA to regulate conventional aircraft noise at America's airports, and the SST would negate these gains. Amid public concern over both the SST and Boeing's new 747 jumbo jet, in 1968 President Johnson signed a noise abatement bill that seemed to preclude the widespread adoption of SSTs at American airports. The SST issue should have been put to bed right there, long before its impact on the atmosphere became a concern.[15]

But the SST was nothing if not resilient. Its economic and environmental liabilities notwithstanding, the plane still had considerable political support—most importantly from Johnson's successor, Richard Nixon. Nixon pushed for the SST for two main reasons. First, Nixon's advisors feared that without an SST, the United States would "take a back seat in aviation history for the next decade."[16] The Johnson administration had invested considerable money in the project, and the Nixon administration hoped to capitalize both economically and politically on that investment. Nixon and his staff expected that international SST routes would expand rapidly, and the market for the planes would expand in turn, giving the nations that manufactured the planes an advantage in their "balance of payments."[17] Second, the administration worried that canceling the SST would draw the ire of American labor unions, with whom Nixon already had a prickly relationship. Nixon's staff was in the process of trying to curb what the president considered "inflationary" wage increases proposed by manufacturing unions, and Nixon had no interest in further alienating workers by cutting high-profile American aerospace jobs during his first few years in office.[18] That those jobs were mostly in Washington State—again, the home of two powerful Democratic senators—made the prospect of terminating the SST even less palatable.

Unlike the Johnson administration, Nixon and his staff actively sought to preempt criticism of the project. The administration released informational fact sheets on the benefits of the SST to American consumers, to the nation's balance of payments, and to the economy in general. In addition, having created (and handpicked members) of the Council on Environmental Quality (CEQ) in August of 1969, Nixon sought to quiet his critics' fears about the environmental effects of SST by incorporating that body's recommendations into the plane's development.[19]

The administration's strategy backfired. Acting head of the CEQ, Russell Train, whom Nixon had tapped as much for his ties to the Republican Party as for his knowledge of and concern for the natural environment, apparently took his job more seriously than the president expected him to. In May of 1970, at hearings convened by Senator William Proxmire of Wisconsin to capitalize on the success of Earth Day, Train summarized the SST's myriad potential environmental problems. He vowed that the administration would put the craft into production only if it could resolve these heretofore irresolvable issues.[20] Compounding the damage, the respected physicist Richard Garwin, who had chaired Nixon's ad hoc Office of Science and Technology Committee on the SST in 1969, also attacked the program, both for its environmental impacts and for its economic liabilities.[21] Train committed to an environmentally acceptable SST, but Garwin concluded that such a plane could not be built. By the time Nixon's 1971 funding request hit the Senate floor in November of 1970, most of the nation's renowned scientists and economists, the CEQ, Nixon's own science advisor, Milton Friedman *and* John Kenneth Galbraith (prominent economists who rarely agreed on anything), perhaps a majority of the American public, and even some inside Boeing all opposed the program.[22] The Senate killed the program's funding by a narrow vote on March 24, 1971.[23]

THE SST AND THE ATMOSPHERE

The first death of the SST, in 1968, is in many ways a case study of grass-roots environmentalism of the 1960s. Public interest driven by backyard concerns, proactive anti-SST campaigns, and political coalition-building combined with common sense to produce a victory for American environmentalists. This familiar story about the SST is particularly instructive because of how it contrasts with the second death of the SST, in 1971. Sonic booms, noise pollution, and economics killed the SST, but the debate over the plane also introduced new players and a new type of global environmental issue to American environmental politics. The players were scientists, and the issue was atmospheric change. Between 1968 and 1971, an increasingly vocal group of atmospheric scientists raised new concerns about the effects of high-elevation supersonic flight on the earth's atmosphere—effects that potentially included ozone depletion, increased solar radiation, and climate change.[24] Scientists at government-funded institutions like NASA and the National Center for Atmospheric Research conducted research into the SST's atmospheric impact in part to fulfill institutional missions to conduct more socially relevant research. But by framing their concerns about the SST's impact on the atmosphere in environmental terms, these scientists also helped to bring CO_2 and atmospheric change into the realm of American environmental politics.

By the late 1960s, NCAR had established itself as the primary locus for atmospheric science in the United States, and it was primed to study the SST. Measuring the plane's effects on atmospheric composition addressed both an institutional mission to study multiple aspects of the earth's atmosphere and growing pressure from the National Science Foundation to demonstrate the practical application of atmospheric and climatic research. As NCAR's administrators knew, growth in atmospheric science in the past decade had benefited the institution, but it had also fostered competition. In 1970, Nixon created the National Oceanic and Atmospheric Administration (NOAA), another major center of atmospheric monitoring, forecasting, and general circulation modeling. NCAR was technically a nongovernmental institution, but it nevertheless relied heavily on federal funding, and the institution's leaders looked upon NOAA warily. Herb Hollman of the Department of Commerce (NOAA's parent organization) fueled these fears. He had no direct authority over the NSF—also a federal agency—but he nevertheless threatened to dislodge NCAR from its NSF moorings and roll it into NOAA unless NCAR made an effort to "pay more attention to the public policy side" of atmospheric research.[25] For NCAR's concerned leaders, SST research provided one way to demonstrate the institution's larger social value.

Studying the SST also fit with NCAR scientists' existing research. During the 1960s, under directors Walt Roberts and John Firor, the institution actively sought to move the atmospheric sciences beyond traditional meteorology. NCAR's scientists hoped to integrate chemical, physical, and numerical expertise to create theoretical and numerical models of atmospheric circulation in the short, medium, and long term at local, regional, and global scales.[26] Both Roberts and Firor believed that integrated atmospheric research at NCAR should serve the "national interest," and NCAR's scientists and managers continued to frame their projects in terms of socially relevant issues like air pollution and weather modification.[27] As early as 1965, the director of NCAR's Laboratory of Atmospheric Sciences, William Kellogg, had taken an interest in monitoring atmospheric ozone using satellites, discussing possible partnerships with NASA or Bell Laboratories to build the necessary instruments.[28] Theoretical concerns about the SST's effects on stratospheric ozone also cropped up among Kellogg's colleagues in the National Academy of Sciences in 1965, and in computer models at the Weather Bureau's Geophysical Fluid Dynamic Laboratory in Princeton around the same time.[29] For Kellogg and other leaders in atmospheric science, the SST ozone problem could not have been more appealing.

Kellogg began to study the SST in earnest in 1968, just as the first wave of environmental concerns over the project began to plateau. Under pressure from Hollman and the NSF to make his basic research more relevant and applicable, Kellogg worked with scientists from institutions around the world to discern whether water vapor, CO_2, and/or nitrogen oxides released by a fleet of five hundred Boeing SSTs could deplete the ozone layer or cause climate change, as some atmospheric scientists—particularly James McDonald of the University of Arizona—feared.[30] In 1968, McDonald explained publicly how, at least theoretically, the SST might increase the water vapor and other gases in the upper atmosphere enough to alter the climate and deplete stratospheric ozone.[31] A decrease in ozone, he warned, could increase the radiation that made its way to earth and thereby significantly increase the risk of skin cancer. While most scientists agreed that a reduction in ozone would lead to an increase in cases of skin cancer, many, including Kellogg, were skeptical of McDonald's claims about the SST and ozone depletion.[32]

In 1970, Kellogg and his colleagues presented their SST research in an MIT-sponsored *Study of Critical Environmental Problems* (SCEP).[33] Despite the paucity of vertical mixing in the upper atmosphere, allowing the gases of the SST's exhaust to stay airborne for between one and three years, SCEP concluded that "no problems should arise from the introduction of carbon dioxide and that the reduction of ozone due to interaction with water vapor or other exhaust gases should be insignificant."[34] But SCEP's authors also recognized that the study had relied on General Electric's potentially problematic data for the company's own engine's emissions, and the report contained many uncertainties. SCEP recommended "that uncertainties about SST contamination and its effects be resolved before large-scale operation of SSTs" began and it outlined a research plan for obtaining better data on the subject.[35]

In August of 1970, before the official publication of SCEP, the *New York Times* and *Los Angeles Times*—both outspoken critics of the SST—ran articles on the report, playing up the recommendation that the project be delayed. The story made the front page of both papers, with the *LA Times* declaring "Scientists Fear Climate Change by SST Pollution" and citing concerns about CO_2 and other gases trapped in the stratosphere. The *LA Times* quoted Kellogg specifically: "When you change something on a global basis," Kellogg told the press, "you had better watch out."[36]

Kellogg's caution was only half the story. He and his SCEP coauthors certainly had concerns about ozone depletion and climate change—Kellogg would express deep concerns about anthropogenic climate change and environmental degradation throughout his life—but based on his research in 1970, he saw no direct evidence that the SST would cause either of these phenomena. In early March of 1971, as the Senate debated the SST's future, Kellogg testified before Congress alongside SST supporter Fred Singer, noting that he had so far found no convincing evidence that the SST would cause significant ozone depletion or climate change.[37] "In short," wrote Kellogg in his prepared statement, "I have found no environmental basis for delaying the Government's SST program."[38] Still, if his testimony gave the forthcoming SST prototype a tentative green light vis-à-vis the atmosphere, Kellogg's cautionary and equivocal SCEP report—along with his comments to the press—hardly provided the plane's backers with a ringing environmental endorsement.

Some scientists worried that Kellogg and Singer had underestimated the SST's impact. Kellogg and Singer agreed that the SST might cause some ozone depletion and could potentially lower the temperature of the stratosphere, but they concluded that these changes would be insignificant compared to natural atmospheric variations. Harold Johnston, a chemist at the University of California, Berkeley, disagreed. He predicted a much greater reduction in ozone due to supersonic flights than SCEP anticipated, if not because of increased water vapor then because of the chemical reactions between nitrogen oxides and the extra oxygen atom of ozone molecules. At a Department of Commerce conference on the SST held in March of 1971 in Boulder, Colorado, Johnston publicly objected to SCEP's conservative analysis of the nitrogen oxide problem.[39] An expert on ground-level ozone chemistry, Johnston claimed that trace gases could significantly reduce the ozone in the stratosphere—enough, he worried, to increase the risk of skin cancer around the world, as McDonald had feared.[40] Though few scientists at the Boulder conference made the jump to Johnston's extreme case of up to a 50 percent reduction in ozone, his concerns raised eyebrows. When he returned to Berkeley, he began writing up his research in a controversial paper that would appear in *Science* in August of 1971.[41]

The first story about the death of the SST, the story about mainstream environmentalism and common sense, does a better job of explaining why Congress killed the SST than does this second story. Johnston's results, leaked to *New York Times* science writer Walter Sullivan in late May of 1971, appeared only after Congress had voted against the SST. But Johnston's work nevertheless found an audience.[42] In Europe, the Anglo-French Concorde was entering the construction phase, and further debates about supersonic transports and their role in American air travel lurked on the horizon. Now the earth's atmosphere—the new wrinkle in the SST story—took center stage. Working with NASA, the Department of Transportation created a four-year atmospheric-monitoring program designed specifically (and belatedly) to study the atmospheric and climatic effects of the SST. The Climate Impacts Assessment Program, or CIAP, vindicated both McDonald and Johnston.[43] The assessment found that a fleet of five hundred Boeing 2707 SSTs would in fact decrease ozone globally by about 12 percent, leading to a potential 24 percent increase in cases of skin cancer in the United States, representing about 120,000 new cases per year, or 2.4 million cases over the twenty-year lifetime of one of the planes. For the Concorde, which flies lower than the Boeing 2707 would have, the numbers were significantly lower, but the nature of the projected impacts were similar. The SST's complex effects not just on ozone but on CO_2, water vapor, and climate still concerned some scientists as well. Amid controversy over the certainty of their results, CIAP's participants recommended further limitations on the Concorde's U.S. flights, based in large part on their effects on the atmospheric environment.

THE ATMOSPHERE AS AN ENVIRONMENTAL ISSUE

Research into the impact of the SST launched scientific and political conversations about a new type of "environmental" problem for the 1970s: the degradation of the earth's atmosphere. Both the New York Times and the LA Times reported on this new kind of pollution. They may have misrepresented the substance of William Kellogg's findings, but their characterization of climatic change and ozone depletion as new environmental issues accurately reflected the commonalities between scientists' and environmentalists' concerns. Kellogg's "you better watch out" warning about the unintended consequences of modifying global systems echoed environmentalists' claims that threats to nature and nature's systems should play a role in decisions about technological and economic development. The SCEP report explicitly underscored this commonality, framing the focus on global environmental problems as a complement to "local and regional problems . . . that many organizations are deeply concerned with studying and ameliorating."[44] In this new context, the curious Keeling Curve no longer described a "grand experiment" or a vaguely troubling trend. Like atmospheric change more broadly, CO_2 became a serious environmental concern.

As Keeling had noted in 1963, CO_2 in the atmosphere was a peculiar type of concern. First, the scale of atmospheric change made it unique among mainstream environmental issues of the 1960s and 1970s. Land-use and waste-management problems typically forced environmentalists to focus on specific partitions of terrestrial or aquatic space, as did efforts to protect wilderness areas and species habitat. Water pollution, though part of dynamic and theoretically global aquatic systems, generally stayed within the confines of watersheds and drainage patterns, and even common air pollution due to industry and agriculture—itself a form of atmospheric problem—was a primarily regional concern. An international environmental movement focused on resources and global environmental degradation grew up in tandem with American environmentalism during the 1960s, but many of the international movement's primary concerns represented no more than a global summation of these key regional issues.[45] Among environmental activists worldwide, NASA's images of the earth from space helped shape an ideal of cooperative environmental protection articulated in the "Only One Earth" theme chosen for the 1972 U.N. Conference on the Human Environment. But when former Sierra Club director and Friends of the Earth founder David Brower famously exhorted environmentally conscious Americans to "think globally, act locally," his "global environment" reflected an amalgamation of "environments" discrete in space and time, each worthy of protection.

The chemical constituents of the stratosphere, which circulated globally in the thin layer of gases encircling the earth, did not fit neatly into this vision. CO_2, ozone, and nitrogen oxides, though measured in parts per million and parts per billion, resided semipermanently as potential environmental offenders in a boundless global space. Traveling at the speed of sound through the gaseous mixture of the atmosphere, the SST might alter the environment in its global totality, and the effects of that alteration might be felt anywhere—or everywhere.[46]

Second, the uncertainties of large-scale, long-term climatic change further distinguished the atmosphere from other aspects of the global environment. The causal link between supersonic transports, ozone reduction, and skin cancer, though controversial, presented environmentalists with an easily identifiable public health concern, albeit one difficult to connect to specific individuals or groups. The human and biological impacts of a change in the temperature of the stratosphere were more difficult to grapple with. The impacts of climatic change on the biosphere might occur over decades, or even centuries. Changes in the atmosphere and climate transcended what the SCEP authors referred to as the "first-order" effects of other large-scale issues like population growth and pollution. Instead, the health of the stratosphere and the stability of the climate constituted an overall "status of the total global environment," with less directly measurable effects.[47] In stark contrast to the sonic booms and airport noise that attracted environmentalists to the anti-SST camp in the 1960s, climatic change as a mainstream environmental issue presented few direct, tangible threats to either human health or the natural world.

Even so, scientists like Kellogg and SCEP organizer Carroll Wilson did not shy away from associating climate research with other problems of the global environment. The *Study of Man's Impact on Climate* (SMIC), for example, a follow-up to SCEP, also sponsored by MIT, positioned humanity's impact on the atmosphere as a mainstream environmental issue, both in content and in presentation. Like SCEP, SMIC's primary recommendations involved science policy, including a $17.5 million global monitoring program. But with an iconic NASA image of the earth from space gracing the report's cover, and a Sanskrit prayer reading "Oh, Mother earth, ocean-girdled and mountain-breasted, pardon me for trampling on you" as the frontispiece, the published study also appealed to the precautionary ethos of the Earth Day teach-ins led by prominent environmental scientists and activists in April of the previous year.[48] By 1971, scientific concerns about the atmosphere had become part of popular concern over the global environment. Atmospheric scientists themselves, however, hardly made typical environmentalists.

SCIENTISTS AS ENVIRONMENTALISTS?

On the Wikipedia page for Charles David Keeling, the term *environmentalist* appears exactly zero times. Neither does *conservationist*, *activist*, or any derivation of the word *political*.[49] Keeling was an outdoorsman amd a longtime member of the Wilderness Society, and in 2005, shortly before his death, he received the Tyler Prize for Environmental Achievement. Keeling's 1963 article on CO_2 as a form of pollution reflected a precautionary ethos that he shared with many self-described conservationists and environmentalists of his era; and his concerns over CO_2, like those of his scientific colleagues, grew out of the cultural milieu that hatched Rachel Carson and David Brower. But like most of his scientific colleagues, Keeling probably would not have described his work on CO_2 and climate change as "environmentalism." Despite the commonalities between scientists like Keeling who were interested in problems of atmospheric change and environmentalists who worried about ecological destruction, the SST debate reveals the tensions that existed between the two groups. These tensions helped define the emerging politics of climate change.

To a large degree, philosophical differences between the two communities revolved around the two groups' attitudes toward technology. Environmentalists embraced scientific developments that might help them forge a more responsible relationship with the world around them, but they objected to the centralization, secrecy, and wastefulness of many government-sponsored big-science projects and the technologies they engendered. To groups like the Sierra Club, the Wilderness Society, and Friends of the Earth, the SST symbolized the worst elements of the aerospace boom of the 1960s.

Stewart Brand's *Whole Earth Catalog,* an oversize review of tools and technologies designed to help readers manage their lives in environmentally sustainable ways, embodied environmentalists' ambivalence about technology.[50] First published in 1968, the *Catalog* embraced the potential benefits of transparently produced, widely disseminated knowledge and technology. For Brand and many other environmentalists, the main problem was not technology itself but its administration by a central authority that did not account for environmental values. Environmentalists were fascinated by new technologies, but they often distrusted the institutions that created them.[51] The *Catalog* enabled individuals to capitalize on technological gains while operating outside of institutional boundaries. Embedded in the *Catalog*'s populist do-it-yourself message was also a critique of the inaccessible, bureaucratically controlled technologies that supported the environmentally irresponsible status quo. Environmentalists were particularly skeptical of large-scale corporate and government science and technology projects like the SST, which they believed ignored citizen input and disproportionately valued technological or material gains over conservation and environmental protection.

Atmospheric scientists were far less distrustful of government. In fact, they relied on the high-tech tools of big-government science to do their jobs. NOAA and NASA were government institutions, and atmospheric scientists at NCAR and elsewhere frequently collaborated in government-sponsored research. The scale of the atmospheric environment required expensive, large-scale research projects. Atmospheric scientists were eager to apply existing and developing aerospace technologies to their research on the global environment, and they were especially excited about the computer and the satellite.[52] They generally sympathized with the conservation goals of the environmental movement, and individual scientists contributed time, money, and expertise to environmental organizations like the Sierra Club and the Wilderness Society. NCAR director John Firor, for example, sat on the board of Environmental Defense.[53] But members of the atmospheric science community often regarded environmental organizations' opposition to the SST as a knee-jerk reaction against technology more generally. Many agreed with Fred Singer when he chastised environmentalists for fostering "a general reaction against all technological progress and against basic science itself."[54] For scientists, the aerospace boom meant bread and butter.

It bears noting that not all scientists were sanguine about all technology. Quite the contrary. Like environmentalists, many American scientists struggled to reconcile their commitments to technological development with their concerns over new technology's potentially destructive impacts. Divisions between scientists often traced disciplinary and institutional lines. Outspoken biological scientists—especially conservation biologists and ecologists working at universities—took the lead in warning citizens about the impacts of technology on the quality of their environment. They opposed developments made possible, and often supported, by chemists, geologists, and nuclear physicists in both government and industry. Rachel Carson's 1962 *Silent Spring,* for example, self-consciously employed the science of ecology to illuminate the environmental risks of cutting-edge chemical pesticides. In the late 1960s, Washington University biologist Barry Commoner and Stanford University entomologist and population biologist Paul Ehrlich each wrote popular books warning of imminent environmental dangers associated with technological development and population growth.[55] As I describe in chapter 4, outspoken climate scientists would try to follow Ehrlich's and Commoner's lead in the 1970s. Some scientists, like William Shurcliff and John Edsall, objected to new technologies in their capacity as citizens rather than as scientists, although activist-scientists like Commoner argued that the two roles should not—in fact could not—be separated.

The Vietnam War also led many scientists to question the impacts of technological development. Politically liberal scientists, primarily at universities, objected to conducting research that would be used in America's military efforts in Vietnam. At MIT, graduate students and some faculty joined protestors who called for a reduction in the more than $43 million per year the institution received from the Department of Defense.[56] On March 4, 1969, a group of graduate students and distinguished professors held a work stoppage to highlight the threat posed to humankind by "the misuse of scientific and technical knowledge."[57] At Stanford, Denis Hayes—later a key coordinator of the first Earth Day, and a character who will reappear in a very different context in chapter 5—helped take over an applied electronics laboratory that conducted classified defense research.[58] According to a *Physics Today* study published in 1972, four out of five physicists surveyed disapproved of conducting classified research on university campuses and opposed Nixon's actions in Vietnam.[59]

Objections to the Vietnam War often entwined with scientists' environmental concerns. The American Association for the Advancement of Science (AAAS), the United States' largest scientific society, passed a series of resolutions between 1965 and 1972 that encouraged a peaceful resolution of the Vietnam War, called for a study of environmental modification techniques used in the war effort, and finally denounced both the use of biological and chemical weapons and the war itself.[60] Scientists' concern for the environment provided a platform from which to criticize other scientists' roles in creating technologies that served the war effort.

The generation of atmospheric scientists that had overseen the tremendous growth of their discipline during the aerospace boom of the 1950s and 1960s—Roger Revelle, Walt Roberts, William Kellogg, and others—genuinely worried about the impacts of technological change on both humans and their environments, but they also understood environmental issues in utilitarian terms. While an FBI file on Roberts from the 1940s and 1950s identified the director as sympathetic to left-leaning causes, atmospheric scientists in 1970 were on the whole less radical than some of their counterparts in particle physics, biology, and environmental science. They saw environmental problems less as fundamental social issues than as scientific challenges. As scientists, they sought to identify the component parts and find solutions. Most atmospheric scientists investigated the supersonic transport's effects on the atmosphere not as a way to undercut the project but as a first step in making the plane better. The second step, they believed, would be to mitigate harmful effects through a better SST design.[61]

If the SST highlighted existing tensions between atmospheric scientists and their counterparts in the American environmental movement, it also forced atmospheric scientists to face tensions over appropriate forms of advocacy within their own ranks. These internal tensions helped shape climate scientists' relationship to environmentalism, typically reinforcing the science-first approach that has since characterized climate change advocacy. In general, even when atmospheric scientists did share environmentalists' views, they had difficulty reconciling direct political activism with their ideals of political neutrality and scientific objectivity. They sought to contain their professional concerns over atmospheric pollution and anthropogenic climate change within the community-defined boundaries of "good science." The "ist" in "scientist" was meant to supersede the "isms" of political or ideological commitments. Ultimately, scientists' commitments to the ideals of good science limited the nature and extent of their environmental advocacy.

The debate over good science is a subtext to many of the stories in this book, and it is worth pausing to explore the key terms of that debate: *objectivity* and *neutrality*. To begin with, the concepts of objectivity and neutrality are not interchangeable. As historian of science Robert Proctor argues, the concept of "value-neutral" or "value-free" science insulates science from political or ideological ends. "Neutrality," he writes, "refers to whether science takes a stand."[62] In Proctor's formulation, neutrality need not have anything to do with objectivity, which he defines as "whether science merits certain claims to reliability."[63] In scientific controversies, however, scientists tend to contest opponents' political neutrality as a way of undercutting their claims to objectivity. Antagonists in scientific disputes play on an underlying assumption that science conducted for a specific purpose or influenced by outside considerations cannot be as reliable as science conducted for its own sake. As a result, the terms—and the concepts—are often conflated.

To complicate matters, neutrality and objectivity are hardly absolute. Most scientists harbor commitments to both ideals, but scientists tend to define these concepts in practice and on the fly, and rarely do they address the terms explicitly except during controversies, if then. Atmospheric scientists studying the SST, for example, rarely used the words *neutrality* or *objectivity* when discussing atmospheric research. But they employed the concepts frequently in their commendations of "good science" and the "good scientist," as well as in their condemnations of politically motivated science and "emotionalism."

During the SST debate, William Kellogg and Fred Singer were at once specialists in the field and arbiters of scientific propriety, and they couched their claims to neutrality and objectivity in terms of good science and the limits of advocacy. Kellogg's review of James McDonald's written testimony on the SST in the spring of 1971, for example, complained that the document was "angry and finger-wagging, clouding the real issues."[64] "He is a good scientist," Kellogg acknowledged, "and most (though not all) of his facts are correct—but he presents the case with too much emotion and a rather transparent hostility to the whole idea of the SST."[65] Kellogg did not immediately object to the details of McDonald's work—in fact, he called for a careful reconsideration of McDonald's main points.[66] Instead, he undermined the validity of McDonald's study by highlighting its apparent bias—a cardinal sin against the ideals of "good" value-neutral science. Scientists like McDonald who criticized the potential environmental impacts of the SST thus expressed their opposition at the risk of damaging their professional reputations as scientists.[67]

McDonald's and Harold Johnston's outspoken objections to the SST may have transcended Kellogg's boundaries of good science, but Kellogg himself frequently used environmental concerns to advocate on behalf of NCAR and the atmospheric sciences more generally within the larger scientific community. Like the concept of neutrality itself, the line between political and scientific advocacy was fuzzy, and institutional leaders often had to balance their commitment to neutrality against the obligation to promote the interests of their science. Funding was a central concern. Through executive-level organizations, the federal government funded the majority of atmospheric science conducted in the United States. NCAR, technically a nongovernmental research center, relied heavily on the federally controlled National Science Foundation. NOAA, a government institution, leaned on its parent agency, the Department of Commerce.[68] NASA was its own federal agency. Each of these agencies, in turn, ultimately depended on Congress and the executive Office of Management and Budget to approve their budgets. Congress increasingly demanded that the nation's scientific institutions directly serve the "national interest," and in the early 1970s, atmospheric scientists were under pressure to make their basic research relevant to social and political issues.[69] Congress hoped to see tax dollars spent on practical solutions to the complex problems of the 1970s. The political salience of environmental issues, along with many scientists' genuine concern for humans' impact on the global atmosphere, prompted some scientists to frame their research on the atmosphere in terms of environmental degradation. In the case of the SST, the trick was to convince Congress and administrators at the NSF and the Department of Commerce that the airplane's potential problems warranted expensive atmospheric research, but without overstating the case and actually undermining the SST—and its budget.[70]

It is easy to dismiss this behavior as "the way science works," but in the larger history of CO_2 and climate change, this sort of framing was more important than it at first appears. In addition to securing funding, institutional leaders like Kellogg sought to ensure that their disciplines remained relevant, both to the scientific community and society at large. The SCEP and SMIC reports show atmospheric scientists promoting the interests of their disciplines. Here is where the secondary debate over "good science" converges with the science-first approach to advocacy. The reports' authors framed global environmental research in terms of a global environmental crisis that required action, but their recommendations revolved almost entirely around the functioning of science itself. The primary audience of the SMIC report, its authors noted, was the scientific community. "Our secondary audience," it continued, "includes those national officials whose decisions govern the allocation of resources to support scientific programs and the directors of those institutes and laboratories where programs must be initiated, expanded, or modified to implement these recommendations."[71]

One of the chief goals of the study was to convince this secondary audience to grant atmospheric and climatic science "a sufficiently high priority to justify allocation of the resources needed"—that is, to present threats to the atmospheric environment as dire enough to merit study.[72] Both SCEP and SMIC emphasized the dangerous uncertainties of interfering with global environmental systems, but their measured conclusions led inevitably to recommendations for more research rather than real domestic public policy changes, at least in the short term.

It is tempting to view this process cynically. But for concerned atmospheric scientists committed to informing good public policy by presenting more and better knowledge, the politics of science funding provided the best—really the only—way to legitimately protect the global atmosphere from degradation. Rarely did scientists pitch their concerns about the atmosphere directly to the public, as environmentalists might, even when their drive for research funding overlapped with genuine environmental concern (as it often did). Environmental advocates appealed to middle-class Americans for support in protecting the nation's wilderness areas and natural resources. Activist scientists appealed to an international governmental elite for a change in high-level science policy in order to monitor and better understand the global problem. If atmospheric science achieved a higher priority in an international scientific agenda, they believed, then the national and international bureaucrats responsible for setting environmental policy would receive better information and therefore make better decisions. The SMIC report's image of the earth and the Sanskrit prayer on its frontispiece may have conjured a popular ethos of environmental protection, but nowhere in either SCEP or SMIC did the authors attempt to mobilize the public at large. For atmospheric scientists, the best environmental advocacy was science advocacy.

AN ATYPICAL ENVIRONMENTALISM

There is little evidence that concerns for the atmosphere played more than a marginal role in Congress's decision to deny funding for the supersonic transport in 1971, but the debate over the ill-fated program introduced large-scale atmospheric change as a mainstream environmental issue for the first time. It also gave atmospheric scientists a greater role in American environmental politics. The global scale of the atmosphere and the complexity of its impacts on natural and human systems distinguished atmospheric and climatic change from other environmental issues. Only atmospheric scientists had the high-tech tools and disciplinary expertise to recognize threats to the earth's atmosphere; and as informational gatekeepers during the SST controversy and afterward, they were forced to act as the primary advocates for this newly threatened environment. Many scientists did not relish this position, but it was one that the climate science community would find itself in repeatedly throughout the history of global warming.

The debate over the atmospheric impacts of the SST established a relationship between climate scientists and environmentalists that would change only slightly until the 1980s. Atmospheric scientists shared with their counterparts at environmental organizations a common precautionary ethos, but scientists made atypical environmentalists at best. Their particular forms of advocacy reflected their professional values. As scientific professionals, atmospheric scientists were meant to value the community-defined ideals of "good science" over their personal political commitments, and they shared a faith in—and dependence on—the technological advances of the aerospace era, which contrasted with many environmentalists' distrust of large-scale high-tech development. Nevertheless, atmospheric scientists took the lead in promoting atmospheric change as a major environmental issue in the 1970s. In particular, they focused on influencing the agenda of international scientific and environmental organizations leading up to the U.N. Conference on the Human Environment in 1972, where scientists would push for a new interpretation of CO_2 and atmospheric change in a new international context.

In part, differences in atmospheric scientists' and professional environmentalists' forms of advocacy reflected differences between the two groups' institutions. Environmental organizations like the Sierra Club and the Wilderness Society operated with a central political mission, supported by members and donors, to advocate for conservation and the protection of nature, wilderness, and the broader environment. They were largely democratic, and they often engaged in collective political campaigns on common issues. They were—and still are—advocacy groups. Scientists, by contrast, worked in competitive—and sometimes competing—institutions, with the primary and overriding objective of producing more and better knowledge. Hierarchies within institutions like NCAR, NASA, and NOAA were chiefly based on merit and seniority, and these institutions answered to bureaucratic agencies. To the extent that the larger atmospheric community had a central collective political motivation, it was the promotion of science itself. Atmospheric scientists framed their research in terms of its potential social applications—applications that included environmental assessments and resource management. When it came to policy, however, the scientific community's first priority was to support programs that allocated more money for scientific research. This, after all, was part of their job.

Markedly absent from atmospheric scientists' appeals for more money and better cooperation was a concerted effort to involve the American public in studying and protecting the global atmosphere. This absence would persist until relatively recently in the history of global warming politics. America's mainstream environmental organizations relied on broad, middle-class grassroots support in their public campaigns to influence domestic environmental policy. Scientists instead sought to gain influence among high-level bureaucrats and government officials at organizations associated with the United Nations, whom they hoped would sponsor the research that would ultimately underpin sound environmental policies. The SCEP and SMIC reports did gain a popular audience, and a related work, *The Limits to Growth,* became an international bestseller thanks in part to its release during the 1972 U.N. Conference on the Human Environment. But while these works certainly had polemical elements, they did not call on the public at large to take action, nor did they call on people to rally around or against specific sectors in business or government.

Atmospheric scientists were themselves largely middle class, and many participated in America's environmental organizations. But as a group they actively sought to divorce their personal values from their professional opinions, and this shaped their approach to the atmospheric environment. They couched their discussions of potential threats to the global atmosphere in the equivocal language of scientific uncertainty, and their calls for action largely targeted government officials and scientific elites who controlled the budgets and agendas of science itself. The scientists believed that more and better science would inform appropriate policy discussions, in which they did not need to take sides.

Scientists' commitments to objectivity and political neutrality continued to limit the extent of their advocacy in the early 1970s. Their cautious scientific concerns about atmospheric change failed to capture the interest of America's major environmental organizations in a significant way until the 1980s, and even then, scientists directed their appeals more toward governments and other scientists than toward the public at large. As a result, scientific concern over changes to the earth's atmosphere and climate initiated by the supersonic transport grew into a distinct—but not altogether separate—form of environmental activism guided more by the professional values of science than by the middle-class consumer values at the heart of mainstream American environmentalism. Only as environmentalism itself became more global and more scientific did atmospheric change become a central concern of America's professional environmentalists.

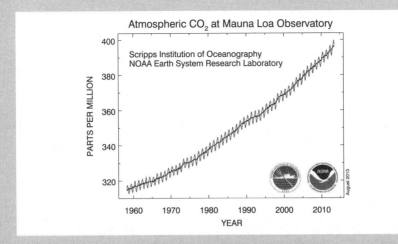

The Keeling Curve—atmospheric CO_2 at Mauna Loa Observatory, 1960–2013. The Keeling Curve is the oscillating line, which tracks the actual monthly average of CO_2 over time. The solid upward slope is the same data corrected for the seasonal cycles that cause the oscillations, which correlates with the annual average of CO_2. Dr. Pieter Tans, NOAA/ESRL (www.esrl.noaa.gov/gmd/ccgg/trends) and Dr. Ralph Keeling, Scripps Institution of Oceanography (http://scrippsco2.ucsd.edu).

Charles David Keeling in the laboratory at the Scripps Institution of Oceanography, 1996. Credit: SIO Photographic Laboratory, Scripps Institution of Oceanography Archives, UC San Diego Library.

Roger Revelle with specimens on the deck of the Scripps Institution of Oceanography research vessel *E. W. Scripps*, ca. 1936. Credit: Eugene Lafond Photographs, Scripps Institution of Oceanography Archives, UC San Diego Library.

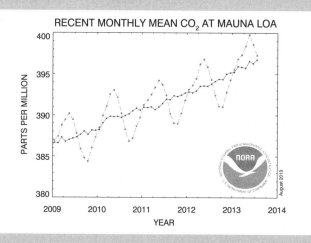

Monthly mean atmospheric CO_2 at Mauna Loa Observatory, 2009–13. The
Keeling Curve is the widely oscillating dashed line, which plots the average
measured CO_2 at Mauna Loa each month, centered in the middle of the month.
The solid through line is the same monthly data, corrected to account for the
seasonal cycles that cause the oscillations. Source: Dr. Pieter Tans, NOAA/
ESRL (www.esrl.noaa.gov/gmd/ccgg/trends) and Dr. Ralph Keeling, Scripps
Institution of Oceanography (http://scrippsco2.ucsd.edu).

Construction of the Mesa Laboratory, Boulder, Colorado, 1965. Credit: Copyright University Corporation for Atmospheric Research, UCAR Digital Image Library.

Walter Orr Roberts (left) and I. M. Pei, Mesa Laboratory, Boulder, Colorado, 1967. NCAR's promotional photographs of the Mesa Lab emphasized the communal spaces the organization hoped would characterize the science conducted at the facility. Credit: "Commemorative Program for NCAR Lab Dedication, May 1967," in *The Design and Construction of the Mesa Laboratory* online exhibit, UCAR/NCAR Archives, www.archives.ucar.edu/search/mesalab/20/10/date_desc. Copyright University Corporation for Atmospheric Research.

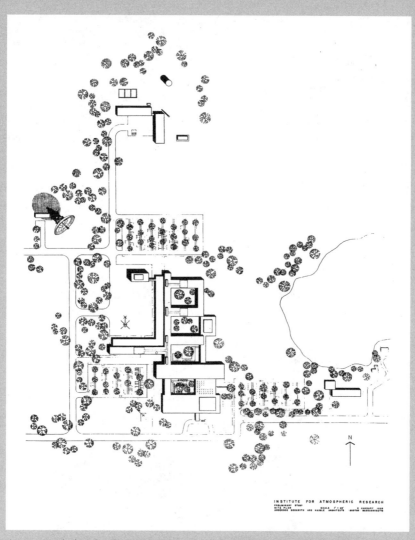

A proposed plan for the National Institute for Atmospheric Research, 1959.
Source: National Science Foundation, *Preliminary Plans for a National Center for
Atmospheric Research: Second Progress Report of the University Committee on Atmospheric Research* (Washington, D.C.: National Science Foundation, Feb. 1959)
(known as the Blue Book), Section IV-26.

Woody Allen at the Mesa Laboratory, Boulder, Colorado, 1973. The Mesa Lab served as the set for outdoor shots of the hospital in Allen's futuristic comedy *Sleeper*. Some NCAR staff, including Stephen Schneider, appeared as extras in the film. Credit: Copyright University Corporation for Atmospheric Research, UCAR Digital Image Library.

An early image of the Boeing 2707 supersonic transport, 1966. Credit: NASA.

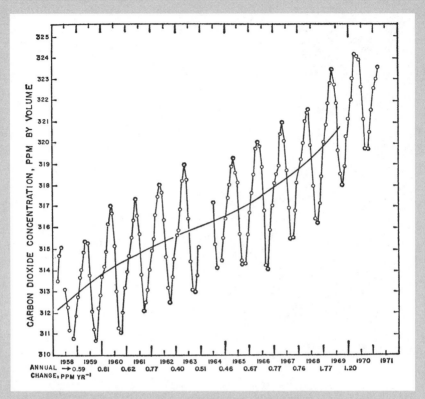

The Keeling Curve, 1951–71. The 1964 gap in the monthly measurements reflects three months when Keeling did not have the funds to continue the Mauna Loa measurements. Source: William H. Matthews, William W. Kellogg, and G. D. Robinson, eds., *Man's Impact on the Climate*, figure 8.11, page 234, © 1971 Massachusetts Institute of Technology, by permission of the MIT Press.

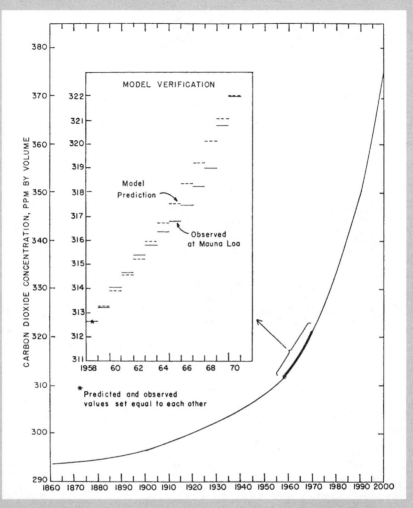

CO$_2$ concentration in the atmosphere, 1860–2000. The *Limits to Growth* study included an image that placed the Keeling Curve (indicated here by Keeling's actual Mauna Loa measurements, 1958–71) in historical and future contexts, showing a long-term trend of exponential growth in atmospheric CO$_2$. Source: Donella H. Meadows, Dennis L. Meadows, Jørgen Randers, and William W. Behrens III, *The Limits to Growth: A Report for the Club of Rome's Project on the Predicament of Mankind* (New York: Universe Books, 1972), 72.

Maurice Strong, secretary-general of the U.N. Conference on the Human Environment in Stockholm, answering a question during a press conference at the Old Parliament Building, June 4, 1972. Credit: U.N. Photo/ Yutaka Nagata.

Maurice Strong, secretary-general of the U.N. Conference on the Human Environment in Stockholm, addressing the Earth Forum's "Whale Celebration," with Stewart Brand (arms crossed) looking on, June 5, 1972. Credit: U.N. Photo/ Yutaka Nagata.

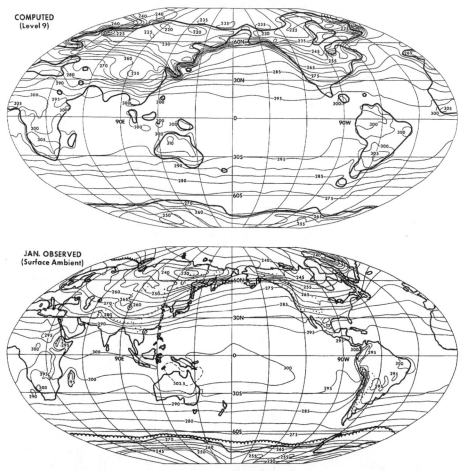

Global distribution of mean temperatures, computed (top) and observed (bottom), 1971. Such world maps enable visualization of temperature data as global. In the 1970s, studying the earth's systems at the global scale helped to reinforce a vision of the global environment that served both scientific and political goals. Source: J. L. Holloway Jr. and S. Manabe, "Simulation of Climate by a Global General Circulation Model: I. Hydrologic Cycle and Heat Balance," *Monthly Weather Review* 99:5 (1971): 344. American Meteorological Society. Used with permission.

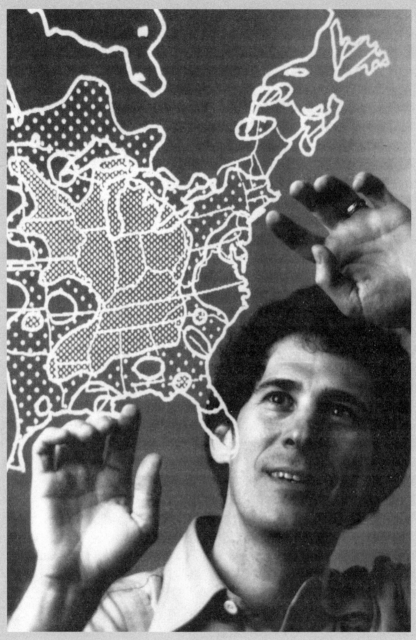

Climate scientist and Intergovernmental Panel on Climate Change lead author Stephen Schneider, ca. 1990. Schneider served on the front lines of the global warming debate, from his controversial 1976 *Genesis Strategy*, through his leading role in the "nuclear winter" debate of the 1990s, to his participation in the IPCC. Credit: Courtesy of American Institute of Physics, Emilio Segrè Visual Archives, Physics Today Collection.

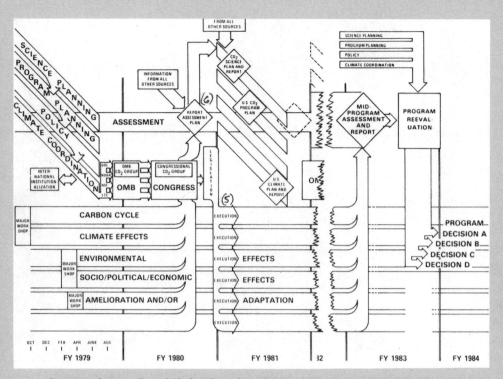

Program director David Slade's flow chart for the DOE-AAAS climate study, 1979–84 (projected). Following the first DOE-AAAS climate workshop in 1979, Slade proposed to integrate CO_2 into the U.S. Department of Energy's research and policy agenda for President Carter's second term (which never came to be). Source: AAAS Climate Program Records, AAAS Archives, Washington, D.C.

NASA climate modeler James Hansen, 1993. In the summer of 1988, Hansen famously told Congress that it was "time to stop waffling" and to say that global warming was real. He persisted in his science-based advocacy up through his retirement from NASA in 2013. Credit: American Geophysical Union, courtesy of American Institute of Physics, Emilio Segrè Visual Archives.

Vice President Al Gore addressing the U.N. General Assembly, June 27, 1997. Credit: United Nations, New York.

3

MAKING THE
GLOBAL ENVIRONMENT

TWO IMAGES OF THE KEELING CURVE PROVIDE A VISUAL UNDERSTAND-
ing of how putting the curve into a new context can change the meaning
of atmospheric CO_2. One image is from the 1971 *Study of Man's Impact on
Climate* and the other was published in the best-selling mainstay of 1970s
environmentalism, *The Limits to Growth*. The SMIC image treats the Keeling Curve in isolation. By now, you know that behind this simple curve lie
several stories—stories about entrepreneurial scientists taking advantage
of the Cold War research system, stories about institutions that reflect
Cold War optimism and anxieties, and stories about "good science" and
the particular forms of environmental advocacy it permitted within the climate science community in the 1960s. The gaps in the curve—the first just
as Keeling began measurements on Mauna Loa in 1957–58, the second when
he lost funding in the spring of 1964—link the stories behind the curve
with the curve itself. Each of these stories in one way or another involves a
reinterpretation of the rise in atmospheric CO_2 and what it means. But the
SMIC image on its own makes no claims about these stories. Insofar as this
is possible, the study provides the basic scientific information necessary to
let the curve speak for itself.

The Limits to Growth, by contrast, self-consciously tells a specific story
about the Keeling Curve. The book's image places the curve in an alarming
new context, projecting the observed data from Mauna Loa forward and

backward in time to demonstrate an exponential increase in atmospheric CO_2 that reinforces the dire environmental predictions of the larger work. *The Limits to Growth* gives CO_2 all of one paragraph in its more than two hundred pages, but when it does it frames rising atmospheric CO_2 as part of a crisis of industrial capitalism, wherein human activities threaten to outstrip the resource capacity of the earth.

Ultimately, though they differed in approach, both studies were parts of a larger effort to reframe CO_2 in light of a threatened global environment—to reinterpret CO_2 in terms of a global scientific infrastructure and an emerging international political framework built to deal with new global problems. The making of this global environment was a complicated process, however. This chapter is about where the study of CO_2 fits into the story of its making.

The seminal event in the making of the global environment as we know it today was the 1972 U.N. Conference on the Human Environment held in Stockholm, Sweden. For scientists, environmentalists, and some U.N. leaders, the Stockholm Conference represented the center of an effort to promote and codify a broader science-based understanding of humanity's global, collective relationship with nature. The "global environment" that scientists and diplomats envisioned addressed the phenomena of the natural world in terms of specific scientific methodologies and political ideals. Many scientists—particularly atmospheric scientists—focused on building more robust and cooperative global networks for scientific research. But scientists' interest in the interdependence of the earth's large-scale systems also implicitly—and sometimes explicitly—meant supporting continuing efforts within the United Nations to foster global political cooperation on environmental and nonenvironmental issues alike. Here was a new form of science-first advocacy, rephrased in explicitly environmental terms and retooled to fit within a specific vision of the global political arena.

The U.N. Conference on the Human Environment was more than just a scientific affair, however; it was also geopolitical event held at a particular moment in Cold War history that reflected the historically and geographically specific environmental, economic, and political concerns of nations and people from around the world. International preparations for the conference made it clear that the politics of the world's threatened global spaces would reflect the concerns of constituencies tied to local, regional, and national geographies rather than the utopian ideals of scientists and well-meaning U.N. bureaucrats.

The difficulties that scientists faced in introducing CO_2 into the framework of international environmental governance developed at Stockholm reflected the familiar intractability of the problem itself, and the conference revealed the limits of the science-first advocacy that scientists had developed to address CO_2 and climate change domestically. These difficulties also reflected the complexities of a changing geopolitical context, a new iteration of the Cold War's influence on the history of global warming. Articulating CO_2 and climate change as an environmental problem presented new and sometimes unexpected challenges in a political context defined by international development and national sovereignty. What made CO_2 appealing as a global scientific problem also made it difficult to incorporate into the politically negotiated vision of global environmental governance developed at Stockholm. In the end, the Stockholm Conference was a challenging place to talk meaningfully about CO_2. Even so, many of the seemingly tangential discussions about global environmental governance that occurred at Stockholm are vital to understanding global warming politics since then. The tortuous politics of the global environment forged at the conference would be woven into the history of CO_2 for decades to come.

CLIMATE, SYSTEMS SCIENCE, AND THE GLOBAL ENVIRONMENT

In its simplest and most literal definition, the term *global environment* is meaninglessly vague. Etymologically, it would be difficult to conceive of two more general words. From the French *environ*, or "to form a ring around, surround, encircle," *environment* simply means "that which environs; the objects or the region surrounding anything." *Global*, meanwhile, from the Latin *globus*, is an adjective describing a round mass, ball, or sphere, although it has also come, after the French usage, to describe a thing or set of things in their comprehensive, all-encompassing, or unified totality. More recently, beginning in the late nineteenth century, *global* began to pertain more specifically to things whole, worldwide, or universal. So in a sense, the global environment is nothing less than the unified and universal totality of all those things that surround or influence, well, everything.[1]

That the U.N. Stockholm Conference endeavored to address the *human* environment provided only the slightest inkling of further specificity. The conference itself was in part an attempt to define this nebulous global environment in a historically specific moment.

The original inspiration for the Stockholm Conference arose out of a Swedish challenge to nuclear development in the U.N. General Assembly in 1967. Inga Thorsson, a Swedish diplomat, and Sverker Åström, Sweden's permanent representative to the United Nations, objected to a proposal for a conference on the peaceful use of atomic energy, on the grounds that it would benefit only a limited number of nations that had nuclear industries. The Swedish delegation suggested an alternative conference that would "focus the interest of member countries on the extremely complex problems related to the human environment."[2] The Swedish government, apprised of the initial proposal after the fact, launched an initiative on the issue within the U.N. Economic and Social Council (ECOSOC) in 1968, proposing that a conference on the human environment be held in 1972.[3] Sweden offered to host the conference in its capital city and to provide a significant financial contribution. ECOSOC, in turn, submitted the proposal to the General Assembly, which unanimously approved the measure on April 2, 1969.[4]

From the beginning, U.N. leaders recognized that any regime of international environmental governance would necessarily include a broad program of scientific research and environmental monitoring. If the earth was, as U.S. ambassador to the United Nations Adlai Stevenson pronounced in 1965, "a little spaceship, dependent on its vulnerable reserves of air and soil," it was up to scientists to find out exactly what kind of shape the ship was in before politicians could make a plan to save it.[5] Scientists, for their part, saw the U.N. Conference on the Human Environment both as a way to create the monitoring infrastructure that would enable study of the environment at a global scale and as a way to make that new global research relevant to the broader world. They anchored their efforts in the most "authentically global" environmental problem they could imagine: the problem of atmospheric change.[6]

Like most U.N. conferences, the Stockholm Conference involved a long run-up to a very short meeting. During that three-year prelude, scientists produced four major collaborative reports on the earth's large-scale environmental systems: the *Study of Critical Environmental Problems* and its follow-up, the *Study of Man's Impact on Climate; Global Environmental*

Monitoring, by the International Council of Scientific Union's Scientific Committee on Problems of the Environment; and the much more famous *The Limits to Growth*, produced by a group called the Club of Rome.[7] These studies were developed more or less concurrently between 1969 and 1972, and they overlapped significantly in subject matter and in personnel. They all advocated a holistic, interdisciplinary "systems science" approach to the global environment that their authors believed would, if implemented, lead to rational and effective environmental policymaking at the international level. With the exception of *The Limits to Growth*, each of these studies was undertaken explicitly for the 1972 U.N. Conference on the Human Environment, and, again with the exception of *Limits to Growth*, each included a significant number of atmospheric scientists and was characterized by an overriding concern for atmospheric and climatic change.

Despite its relative reticence on climate, *The Limits to Growth* provided the clearest and most popular application of the systems science methodology behind all four efforts. A general term, *systems science* describes a multifaceted movement within the scientific community after the Second World War to use digital computers to numerically simulate complex, nonlinear phenomena.[8] *The Limits to Growth* built upon a subfield of systems science called systems dynamics, a method of studying complex organizations over time pioneered by computer engineer Jay Forrester, who became a management science expert.[9] *The Limits to Growth*, published in the spring of 1972, grew out of a project initiated by an elite international group of scientists, industrialists, businesspeople, diplomats, and leaders of civil society founded by Italian industrialist Aurelio Peccei, called the Club of Rome. Drawing on Forrester's "World 1" and "World 2" computer models, the club focused on the genesis of and possible solutions to humanity's common and persistent problems: "poverty in the midst of plenty; degradation of the environment; loss of faith in institutions; uncontrolled urban spread; insecurity of employment; alienation of youth; rejection of traditional values; and inflation and other monetary and economic disruptions."[10] Global in scope, deeply interrelated, and changing at an ever-accelerating rate, these issues constituted what Club member Hasan Ozbekhan defined as a "generalized meta-problem (or *meta-system of problems*) which we have called and shall continue to call the 'problematique.'"[11] The Club thus came to call this collection of issues the *monde problèmatique*.

As the name of the study suggests, the *Limits to Growth* group focused on the functional boundaries—or limits—of the world's interrelated large-scale systems. Funded by the Volkswagen Foundation and led by one of Forrester's students, Dennis Meadows, the *Limits* group set out to examine the "five basic factors that determine, and therefore, ultimately limit, growth on this planet—population, agricultural production, natural resources, industrial production, and pollution."[12] Meadows and company used a computer model very similar to Forrester's World 2 to demonstrate the liabilities of continuous demographic, economic, industrial, and agricultural expansion.[13] The world, the Meadows group argued, would soon be "faced with an inevitable transition from world-wide growth to global ecological equilibrium."[14] Without a large-scale collective effort to curb growth intentionally, this transition would occur on its own—abruptly and catastrophically—sometime in the coming century. Released a month before the Stockholm Conference, the published book sold more than seven million copies in thirty languages, and it became a mainstay of popular environmentalism, both in the United States and abroad.[15]

The Limits to Growth described geographically discrete problems like soil depletion, mineral resource shortages, pesticide use, air and water pollution, overfishing, population growth, and nuclear waste in terms of a single, interdependent world system, but the study paid only cursory attention to the global atmosphere.[16] The Keeling Curve set within an exponential curve of atmospheric CO_2 rise was apparently all the book's authors thought they needed to reinterpret CO_2 in terms of the *monde problèmatique*. That was not the end of their interest in CO_2, however. Concurrent with their work on *The Limits to Growth*, several Club of Rome members and consultants took a keen interest in problems of the atmosphere. In particular, Carroll Wilson of MIT's Sloan School of Management, former chair of the Atomic Energy Commission, recognized that atmospheric scientists' previous success in using models to analyze and predict the behavior of nonlinear feedback loops operating at local, regional, and global levels helped validate the Club of Rome's holistic approach. What systems dynamicists sought to do with the earth's environmental, social, economic, and political systems more generally, atmospheric scientists had already begun to accomplish with the earth's climate. Weather forecasting, after all, was the first use to which the first computers were put. Atmospheric science was the original systems science.[17] Between 1970 and 1972, while *The Limits to Growth* was still in gestation, Wilson teamed up with atmospheric scientists, climatologists, and other management experts to incorporate the atmosphere—and by association, climate change—into a scientific vision of the threatened global environment.

The first and most important study to address climate as a major environmental issue was the 1970 *Study of Critical Environmental Problems*—the same study that addressed the atmospheric impacts of the supersonic transport. Principally organized by Wilson, SCEP involved a veritable Who's Who of atmospheric science, complemented by a smattering of notable management professionals (mostly from MIT), biologists, and scientific administrators from academia, government, and the private sector.[18] Wilson and his eleven-person steering committee hoped to "fill the gap" in research into "global problems such as changes in climate and in ocean and terrestrial ecosystems" that "had not been subjected to intensive study and examination." The resulting study synthesized more than two hundred papers and scientific articles into a first assessment of "the status of the total global environment."[19]

The SCEP group actually assessed two things. First, through working groups on the climatic and ecological impacts of human activities, on the implications of environmental change and remedial actions, and on agricultural waste and energy production, the group outlined the major threats to the global environment for the 1970s. The climatic effects of CO_2, airborne particulates, and SST emissions figured prominently in this discussion, and the published SCEP included one of the first widely distributed images of the Keeling Curve available to the public. But the SCEP also dealt with DDT and other pesticides, mercury and other heavy metals, nuclear waste, more "traditional" sources of air and water pollution like agriculture and industry, and, finally, oil spills.[20] The SCEP report was in this sense a fairly comprehensive "state of the problem" assessment of the global environment—a sort of tentative, environment-specific preview of *The Limits to Growth*.

The SCEP authors were just as concerned with the effort to study the global environment as they were with the environment itself, however. The global environmental crisis, like the specific issue of CO_2, was for the SCEP authors first and foremost a scientific problem. In addition to outlining the global environmental crisis, SCEP offered a detailed assessment of large-scale, systems-based environmental science.[21] Steeped in the belief that better science would help make better policy, the SCEP group focused their suggestions almost entirely on making more and better science. In general, they recommended a more extensive and cooperative international environmental monitoring effort, better data collection, and the development of rigorous, universal standards of observation and data analysis. Each working group also offered recommendations specific to research in their particular areas of expertise. For SCEP scientists, these programmatic recommendations were meant to set the parameters for understanding the global environment in the years to come.

SCEP's assessment of the state of global environmental science paralleled an effort by the International Council of Scientific Unions to incorporate similar recommendations into a scientific infrastructure under the auspices of the United Nations. In 1971, the ICSU's Scientific Committee on Problems of the Environment (SCOPE) presented a global scientific research plan to the United Nations that would use the World Meteorological Organization's Global Atmospheric Research Program (GARP) to begin answering the scientific questions raised by the SCEP report. SCOPE published the plan as *Global Environmental Monitoring*, a preparatory document for the 1972 Stockholm Conference.

Climate change provided a centerpiece for both the SCEP and SCOPE reports, and interest in problems of CO_2 and atmospheric change continued to increase in the run-up to Stockholm. The flourishing of systems science acronyms did not end with SCOPE, and a number of other studies honed in on climate as the "most relevant" issue in the project of global environmental monitoring.[22] In 1971, the problem of climate engendered its own international study, a climate-specific follow-up to SCEP called the *Study of Man's Impact on Climate*—the study that produced the Keeling Curve image. Sponsored by MIT and hosted by the Swedish Academy of Sciences, the conference that produced the SMIC report endeavored to forge a consensus on "what we [scientists] know and do not know" about climate and "how to fill the gaps."[23] The report, published in a rush in 1971, offered suggestions for improving climatic and atmospheric research within the international scientific community, including the potential costs of the recommended research programs. Targeted at providing "input into planning for the 1972 U.N. Conference on the Human Environment," SMIC planted the climate and the atmosphere—or, more accurately, climatic and atmospheric science—firmly at the center of its definition of the global environment.[24]

The scientists involved in the SCEP, SMIC, and SCOPE reports built their image of the global environment around the issue of climate change for several reasons. Most obviously, many of the prominent individuals who participated in the studies were themselves atmospheric scientists or climatologists, and both the practical problems and professional ambitions of their disciplines guided their studies. Stephen Schneider, the so-called rapateur in charge of scholarly communication for the conference that produced SMIC, recalled that promoting international efforts to study the global atmosphere served atmospheric scientists' professional self-interest. "Atmospheric scientists had a common cause in all countries," Schneider remembered. "We needed expensive satellites, balloons, ships, and computers from our governments to do our work. International cooperation in data sharing reduced costs of these tools to individual nations."[25] When atmospheric scientists described the global environment in the early 1970s, it was no coincidence that their global vision matched their global research interests.

The SCEP and SMIC scientists' focus on climate went beyond professional self-interest, however. Just as importantly, climate served as an intellectual anchor for the holistic, global-scale approach to environmental research increasingly advocated not only by atmospheric scientists, but also by biologists, oceanographers, and management professionals. Bent on providing input into the U.N. Conference on the Human Environment, SCEP sought out "environmental problems whose cumulative effects on ecological systems are so large and prevalent that they have worldwide significance."[26] As SCEP pointed out, pesticide use, agricultural waste, and population growth were prevalent issues in states, nations, and regions throughout the world. These problems were endemic, and thus in the aggregate they were "global" problems. Still, these issues all had immediate causes and direct impacts, and they were typically rooted in discrete and independent geographical spaces.

The very concept of climate, by contrast—again, derived from the Greek *klima*, or inclination, as in the inclination of the sun vis-à-vis points on the earth—implied a supraregional causal force beyond individual humans' immediate perceptions. Climate change was not only a common problem; it was a problem determined by the chemical composition of common space: the atmosphere. CO_2 and other chemical constituents circulated almost limitlessly in this ubiquitous mixture. As scientists like Stephen Schneider and others had by then begun to show, the potential impacts of climate change on the earth's natural and human systems were similarly limitless. Tied to the global atmosphere—fluid, dynamic, and borderless—climate change offered a prime example of an environmental problem that could properly be addressed only at a global scale.[27] In the early 1970s, the very peculiarities of climate change that had made CO_2 a problematic pollutant for Charles David Keeling in 1963 suddenly seemed like assets for framing CO_2 in a new global context.

THE POLITICS OF SYSTEMS SCIENCE

In the early 1970s, systems scientists sought to trace the Keeling Curve onto a new background of globally focused systems science, but as they did so they also traced CO_2 onto a new geopolitical background. The systems science concepts at the heart of the SCEP, SMIC, and SCOPE reports served a mixture of scientific and political ends. Through these studies, scientists advanced a framework for studying the earth's threatened landscapes, natural resources, and species. By setting the research parameters for measuring and monitoring the world's large-scale systems, they theoretically made the global environment knowable. But each of these studies, written as preparatory documents for the 1972 U.N. Conference on the Human Environment, also engaged in an ongoing political discussion about how the global environment—and by extension the world more broadly—ought to be governed. Global environmental research and global environmental governance were two sides of the same coin.

The political implications of systems science represented a new, environmentally focused take on what nuclear biophysicist and *Bulletin of the Atomic Scientists* cofounder Eugene Rabinowitch had in 1946 dubbed "scientific internationalism."[28] With roots in early Cold War debates about secrecy in nuclear energy research, scientific internationalism comprised two complementary beliefs. The first was that science flourishes best in open, democratic societies that foster the free flow of scientific information across political and geographical boundaries. The second was that cooperation among governments in scientific and technological ventures would foster political cooperation, peace, and global prosperity.

As climate science historian Paul Edwards explains, atmospheric scientists who relied on global networks of meteorological and climatic data had a particular interest in international scientific cooperation. Without organizations like the World Meteorological Organization, they would not have the data to make their science possible. In 1972, however, the politics of systems science went beyond the long-standing support for what Edwards calls "making global data."[29] Systems scientists presented the global environment in a way that all but demanded a paradigm of cooperative global governance outside of the realm of science. The borderless, interdependent global systems that scientists described contrasted sharply with the competitive, state-centered politics of the Cold War.

Indeed, the leader of the U.N. Conference on the Human Environment, Maurice Strong, and many of his U.N. colleagues hoped that the conference in Stockholm would mark a transition away from divisive East-West politics toward a "new globalism" that revolved around a robust and democratic system of international politics. When scientists like Dennis and Donella Meadows and Carroll Wilson described the earth as a deeply interdependent whole, they supported this global vision. In an interconnected world, actions by any one nation or even one industry affected the world's predicament as a whole. Conversely, the earth's systemic limits applied to all nations and groups in all circumstances. Cooperation represented the only "rational" response to "a set of decisions that could govern the future habitability of our planet."[30] Borders, key points of control for the globalizing forces of trade, migration, and disease, had little meaning in the face of transnational environmental problems like air pollution, extranational problems like ocean dumping and fisheries depletion, and truly global problems like ozone depletion and climate change. "The world" as imagined by the graphs and tables in The Limits to Growth had no borders. In fact, the world had no political geography at all. As U.N. Secretary-General U Thant suggested in his foreword to The Limits to Growth, unilateral decisions, belligerent nationalism, and Cold War posturing had no place in a finite world composed of deeply integrated social, political, and environmental consequences.[31] SCEP, SMIC, and The Limits to Growth thus backed up the "Only One Earth" slogan of the upcoming U.N. Conference on the Human Environment with a "we're all in this together" form of science.

"A NEW KIND OF GLOBALISM"

Built upon global environmental concerns that included CO_2, scientists' vision of the global environment implicitly supported cooperative forms of international governance. Maurice Strong, the U.N. point man for the Stockholm Conference, supported international governance more explicitly, but his political goals had little to do with CO_2. Though Strong himself consistently demonstrated concern for scientific research in the service of environmental protection, the "global vision" that shaped his approach to Stockholm reflected geopolitical values that had as much to do with the Cold War as they did with the so-called global environmental crisis. Strong and his colleagues saw in the pressing but poorly defined environmental crisis the type of collective, international problem that the now ailing United Nations had originally been designed to address. Strong hoped to use the Stockholm Conference to re-empower the United Nations as a force for international peace and cooperation by mobilizing its institutional machinery to deal with the world's environmental problems.[32] His concerns underscored the complexities of the geopolitical context into which the Keeling Curve had meandered in the early 1970s.

Strong took the job of conference secretary-general in January of 1971 as a rising star within the United Nations. That same month, the man who appointed him, U Thant, announced that he would not serve a third term as U.N. secretary-general, and Strong became a leading candidate for the job. Strong was a rags-to-riches Canadian businessman whose "disarming modesty and beguiling punctiliousness," according to the *New York Times*, belied a "phenomenal dynamism and an extraordinary skill at both business and diplomacy."[33] He cut an unimposing figure—his naturally slight build had softened in middle age and his thick, close-cut sideburns and short-bristled mustache did little to compensate for a receding hairline—but his presence made an impression. A former president of one of Canada's biggest utility companies, the forty-two-year-old's extensive business experience and his myriad contacts in the oil and gas industries made him an appealing candidate to U.N. members concerned about international development. His success as the head of Canada's international aid program—which grew from $80 million to $400 million under his directorship—complemented his business résumé.[34]

Strong envisioned the U.N. Conference on the Human Environment as the centerpiece of a larger effort to promote global international cooperation, democracy, and unity within the hierarchy of U.N. values. Central to this was the strengthening of the General Assembly, where the United Nations dealt with cultural, social, and now environmental concerns. In 1970, real power within the United Nations was divided unequally among three main groups: the five permanent members of the U.N. Security Council (France, the United Kingdom, China, the United States, and the Soviet Union); the so-called Bretton Woods institutions, a set of ailing international financial institutions responsible for financing development and maintaining stable monetary exchange rates; and the all-inclusive General Assembly.[35] Despite its size, the General Assembly was the least powerful of the three. Like many of his colleagues from "third-party" countries in Scandinavia, South America, and parts of Asia, Strong objected to the U.N. Security Council's dominance of U.N. politics. Nations with no nuclear weapons and no representatives on the Security Council criticized this small governing body as undemocratic. They feared that the superpowers' central concern over "peacekeeping" and international security undermined the other functions of the United Nations.

Strong presented the U.N. Conference on the Human Environment as a way to strengthen the General Assembly and revitalize the "soft side" of the U.N. mission at a critical moment in international relations.[36] The crisis of the global environment—a crisis made up of interrelated threats to the air, water, land, and people of nations around the world—offered the United Nations an opportunity to transcend East-West political tensions by working toward the common objective of environmental protection. "If the member governments of the United Nations can find the political will to do so," Strong proposed, "they can make the conference on the human environment the starting point of a world mobilization for the future of man."[37] Cooperation, he argued, lay at the very foundation of the United Nations' mission. And so too, he suggested (echoing the systems science creed), did cooperation underpin a healthy environment. "The entire global system on which all life depends," he pronounced in 1971, "must inevitably and inexorably lead us back to a new kind of globalism."[38]

DEFINING THE ENVIRONMENTAL CRISIS

Strong's "new globalism"—a vision of cooperative international politics led by a reinvigorated, more democratic United Nations focused, at least at first, on environmental issues—meshed almost seamlessly with systems scientists' holistic image of a fragile, solitary earth made up of complex and deeply interconnected environmental systems. As the Stockholm

Conference approached, however, this new global vision ran headlong into the realities of an international political system built upon regional and national self-interest. Within this competitive system, the rubric for what constituted an important environmental issue was hotly contested. U.N. member states sought to promulgate their national economic and political objectives by recasting the key problems of this new global space in terms of their preexisting political concerns. In contrast to the modeled problems of "the world" presented by *The Limits to Growth*, these real problems had an important political geography. The 1972 Stockholm Conference soon became as much a competitive struggle to define the global environment as it was a cooperative effort to protect it. And in a pattern that would persist throughout the rest of the century and into the next, that battle would revolve primarily around the relationship between environment and development.

That battle began with the inception of the conference itself. The original proposal for the conference in 1968 reflected a growing international interest in the middle-class quality-of-life issues that also lay at the core of mainstream American environmentalism.[39] Europeans watched as their rivers caught fire and their forests died.[40] Industry polluted their air and their water.[41] In concert with growing political support for international cooperation and nuclear disarmament, the "progressive" nations of Europe began to value humans' interactions with the natural world as part of a healthy society. "Environmental" problems were at once natural and social. They included not only the destruction of wilderness and open space, waste disposal problems, and air and noise pollution but also uncontrolled urban growth and increased traffic congestion and accident rates. Environmental degradation, conference promoter Sverker Åström argued, caused "problems connected with family disorganization, mental tensions, and increased crime rates."[42] U.N. Secretary-General Thant worried that an exponentially increasing population would stress ecosystems and natural resources and, in turn, would undermine the stability of modern societies.[43]

The problems of the global environment that Åström and Thant articulated failed to resonate with the developing world, however. At a twenty-seven-nation preparatory meeting in New York in 1970, the self-described less developed countries (LDCs) noted that the United States, the Soviet Union, and Europe bore primary responsibility for problems like industrial pollution, uncontrolled urban development, and toxic-waste management. These issues, alongside concerns about wilderness and open space, also disproportionately affected highly developed, industrialized nations. In short, the LDCs complained, the chief environmental problems identified for the Stockholm Conference were problems of economic success.

If the industrialized world understood the global environmental crisis as a consequence of unchecked industrial development, the LDCs, who had yet to experience industrialization, focused instead on the environmental consequences of underdevelopment. Unlike the "quality-of-life" issues facing the developed world, the LDCs' concerns over poor water, poor sanitation, poor nutrition, and poor public health involved the maintenance of life itself.[44] These were the problems of poverty. LDCs worried that international environmental regulations would exacerbate these environmental problems by upsetting trade and stifling development. Leaders from Latin American nations and the recently independent countries of Africa had little interest in participating in a conference focused on First World concerns about urban growth, recreational open spaces, endangered species, and industrial air pollution. They were even less keen if the resulting environmental regulations might interfere with their own economic development. Some suggested cynically that environmental concerns "were a neat excuse for industrialized nations to pull the ladder up behind them."[45]

The LDCs had a morally compelling argument. For once, they also had a compelling political advantage. Leaders in the developed world had by 1970 already committed to using the United Nations to create a framework for international environmental regulation, and these leaders realized that such an effort would succeed only with the participation of the developing world. LDCs were the majority of U.N. members and produced many of the material resources potentially subject to regulation. In a purportedly democratic United Nations, the human environment that U.N. leaders like Strong and Thant sought to outline would have to reflect visions of the environment from both sides of the industrial divide.

In June of 1971, a Panel of Experts on Development and Environment met at Founex, Switzerland, to "consider the protection and improvement of the environment in the context of the urgent need of the developing countries for development."[46] Worried that some LDCs might choose not to participate in the planned 1972 conference, the informal panel sought to reassure the developing world that their interests would be represented in Stockholm. The meeting articulated the concerns of the developing world and produced one of the foundational documents of contemporary international environmental politics, the "Founex Report."

The meeting at Founex reminded observers that any regime of global environmental protection required global political buy-in, and that buy-in hinged on the traditional national and regional interests of individual nations. The term *global* referred as much to the geopolitical interests at stake as it did to a single unified geographical space. The developing world was ready to commit to a common global effort to protect the environment, but only if "the environment" could be made meaningful to a developing-world political constituency overwhelmingly concerned with issues of economic and social advancement.[47]

Not surprisingly, problems of the global atmosphere did not readily appeal to a developing-world constituency. In fact, as dean emeritus of American foreign policy George F. Kennan noted, for truly geographically global problems like atmospheric and climatic change, "subject to the sovereign authority of no national government," there was virtually no committed political constituency at all.[48] The problems involved in interpreting CO_2 as a form of pollutant that Charles David Keeling had articulated in 1963 persisted in the new international context of the U.N. Conference on the Human Environment. Even among industrialized nations, the vague and geographically boundless problems of the global commons took a back seat to the tangible, geographically specific problems that drove local and national environmental campaigns. Central as it had been to defining the global environment scientifically, climate change thus played only a marginal role in defining the global environment politically. And yet, the politics of the global environment negotiated at Stockholm would shape the landscape of climate change politics for decades to come, and the proceedings themselves would shed light on the quirks of that terrain.

THE U.N. CONFERENCE ON THE HUMAN ENVIRONMENT

The U.N. Conference on the Human Environment was the international community's first attempt to deal with the environment as a meaningful global political issue, and it unfolded as a mix of the deadly serious and the patently absurd. A bizarre episode captured by *New York Times* science writer Walter Sullivan is telling.[49] In a field on the outskirts of Stockholm, the United States' most prominent hippie group, the Hog Farm Commune (founded by Wavy Gravy), hosted not a sit-in or a teach-in or a hunger strike but a self-described "whale-in," protesting whale hunting and supporting an international whaling moratorium. The event, which entailed what Sullivan described as "the lugubrious cry of whales across the pine-studded landscape," generated by both tape recordings of actual whales and human imitations, drew a surprising assortment of conference delegates and government officials.[50] Maurice Strong himself gave a speech supporting a whaling moratorium, which was followed by a brief poem, "ostensibly written by a whale, with prolonged moans and groans." Stuart Brand, the flamboyant acid-test veteran of *Whole Earth Catalog* fame, then mounted the stage in a "plumed top hat" to introduce former secretary of the interior and occasional bugbear of environmentalists Walter Hickel, who gave another speech. The next day, the Hog Farm hippies used plastic bags to dress their bus up as a whale and then followed it through the streets of Stockholm, making whale noises and shouting the Swedish word for whale, "val." In the end, the conference passed a serious-minded whaling accord, a compromise agreement resulting from hard-nosed negotiations among high-level diplomats.

The circus notwithstanding, the framework for environmental governance established at Stockholm has guided discussions about the global environment for four decades. When the conference finally got underway after three years of careful preparation in June of 1972, it involved two types of discussions about the environment. The first kind revolved around establishing practical international mechanisms to monitor and control environmental degradation, and here, the United States took the lead. Emphasizing institutional coordination, cooperative scientific research, and a few key environmental issues that resonated with President Nixon's domestic constituents, the U.S. delegation introduced six limited but pragmatic initiatives for building an international environmental bureaucracy much like what Nixon had established at home.[51] As the president had already learned, nothing could take the air out of 1960s-style environmental idealism better than a 1970s-style environmental bureaucracy.

Despite a dearth of professional scientists in the delegation, the administration's proposals included significant support for international environmental science—in particular, for research into the global atmosphere. In Stockholm, U.S. delegation leader and Council on Environmental Quality chairman Russell Train and his colleagues unveiled Earthwatch, an international program of scientific cooperation intended to provide a framework for assessing environmental problems on a global scale. Earthwatch had two main objectives. The first was essentially bureaucratic. Much as the National Oceanic and Atmospheric Association—created by Nixon in 1970—had streamlined redundant state and federal research efforts at home, the U.S. proposed that Earthwatch coordinate, expand, and reorganize existing national and international environmental research efforts at the United Nations. The second objective was programmatic. It involved launching collaborative efforts to "measure trends and identify problems requiring international action."[52] In particular, Earthwatch focused on the potential threats to the oceans and the atmosphere that the SCEP and SMIC studies had identified. Robert White, director of NOAA and a member of the U.S. delegation, identified the accumulation of dust particles, a rise in atmospheric CO_2, and a decline in atmospheric ozone as three of the program's primary concerns.[53] The World Meteorological Organization, a specialized U.N. agency, would take the lead in this new research.

On the surface, Earthwatch looked much like the type of large-scale global environmental research that systems scientists had called for in the SMIC, SCEP, and SCOPE reports. For many atmospheric scientists, Robert White was a welcome late addition to the U.S. delegation, and his emphasis on climate matched their own interest in the subject. And as a coordinating unit, Earthwatch initially managed to "'integrate' information gathered from across the U.N. system" relatively effectively.[54]

But it quickly became clear that Earthwatch was more of a bureaucratic clearinghouse for existing research efforts than a program to develop the sophisticated infrastructure for scientific research called for by the atmospheric science community. The SCEP, SMIC and SCOPE studies proposed broadening and deepening global scientific research through a creative new systems science methodology. Earthwatch merely sought to integrate existing research—albeit worthwhile research—already being conducted "within the U.N. system."[55] It was, at best, a limited offering, emblematic of what one scientist described as a "lack of imagination" in the Nixon administration's approach to environmental research.[56]

The administration's cautious, bureaucratic approach extended to the delegation's major financial initiatives as well. The U.S. delegation's most important proposal at the Stockholm Conference was a voluntary five-year plan to raise $100 million for U.N. environmental programs, called the U.N. Environment Fund. The administration designed the fund primarily to support research, education, and environmental monitoring within the international community, but the money was also meant to strengthen individual nations' environmental management capabilities through regional-level training and coordination. The administration, however, specifically cautioned against allowing the fund to be diverted into intranational economic or even environmental development programs.[57] The fund would provide the financial and bureaucratic support necessary for implementing Earthwatch, along with the U.S. delegation's other institutional initiatives. But in a move that presaged later administrations' efforts to separate financial support for environmental goals from support for development, Train and his colleagues segregated the U.N. Environment Fund from geopolitical concerns over development and international aid.

In general, Nixon and his aides sought to streamline and publicize the United Nations' existing American-sponsored environmental programs, but the president had no interest in empowering the body by creating new or autonomous international agencies. Nowhere was this more evident than in discussions about the management of the U.N. Environment Fund. The fund both required and supported some sort of bureaucratic structure within the United Nations, and the U.S. delegation's proposal for a "small coordinating unit" mirrored efforts to coordinate and streamline the environmental bureaucracy at home.[58] Within the United Nations, a variety of agencies, including the Food and Agriculture Organization, the World Meteorological Organization, and the World Health Organization, worked on aspects of environmental monitoring, protection, and management. A number of conference delegates thought that the United Nations should create an entirely new agency to deal with environmental problems. The Nixon administration disagreed, fearing that a separate agency would be costly and redundant. Instead, the United States favored a permanent environmental secretariat that would coordinate existing agencies' environmental activities, much as the domestic Council on Environmental Quality coordinated the environmental activities of U.S. federal and state agencies. The United States held the lion's share of the U.N.

Environmental Fund's purse strings, and the conference eventually backed the plan and created the U.N. Environment Programme, an environmental secretariat based in Nairobi, Kenya, with Maurice Strong as its head.[59]

The U.S. delegation also supported three international initiatives that resonated with environmentalists at home: a convention on ocean dumping, a ten-year moratorium on whaling, and the establishment of World Heritage Sites, modeled after the American National Park System. Each of these international initiatives spoke to widely held domestic concerns; and alongside Earthwatch and the U.N. Environment Fund, discussions about these issues at Stockholm enabled the U.S. delegation to capitalize politically on America's environmental leadership at home and abroad.[60]

In Stockholm, however, the U.S. delegation's achievements were quickly overshadowed by a second set of discussions about the global environment that the Nixon administration had explicitly hoped to avoid. Initiated by Sweden and China and then taken up by developing nations, these were philosophical and diplomatic discussions that put environmental protection in the context of development, nuclear disarmament, and international peace. Much as Nixon's aides attempted to address environmental issues in terms favorable to broader U.S. domestic and foreign policies, so too did U.N. members seek to redefine "the environment" to fit their own economic and political interests.[61] And just as Nixon had feared, these interests frequently collided with the geopolitical priorities of the United States.

For the most part, the geopolitical concerns associated with the global environment were already on the table when the delegates arrived in Stockholm. Four preparatory meetings in 1971 and 1972 brought most of the main issues to the fore long before the conference began, and drafts of the "Declaration on the Human Environment," slated for release at the close of the conference, were also largely complete. The Founex Report had made it clear that the relationship between environmental protection and Third World development would be a central concern of many conference participants. Negotiations over language dealing with nuclear weapons and weapons testing in the spring of 1972 presaged a discussion about disarmament in Stockholm, and tensions between "superpowers," nonnuclear nations like Japan and Sweden, and the less developed countries of South America, Africa, and Asia promised to pervade discussions about international environmental protection.

For the U.S. delegation, however, there were a few important and somewhat unexpected wrinkles in Stockholm that undermined the administration's public relations goals for the conference. First, despite pointed and repeated attacks on the administration's approach in the months leading up to Stockholm by the president's domestic opponents, the Nixon administration seems to have been unprepared for the extent to which nations critical of American foreign policy used the environment as a way to attack the United States politically in the official conference forum. Just as at home, the main issue in Stockholm was Vietnam. During his opening remarks on the first day of the conference, Swedish prime minister Olof Palme charged the United States with engaging in "ecological warfare" in Southeast Asia. "The immense destruction brought about by indiscriminate bombing, by large-scale use of bulldozers and herbicides is an outrage sometimes described as ecocide," he declared, echoing statements made by Claiborne Pell on the Senate floor in the previous months.[62] Palme framed the United States' destruction of the Vietnamese environment in terms of a larger relationship between armaments and environmental destruction. He called on the conference to urge an end to large-scale arms production worldwide. The Chinese delegation put an even finer point on the Swedes' criticism. Tang Ke, head of the delegation, asked the conference to "strongly condemn the United States for their wanton bombings and shellings, use of chemical weapons, massacre of the people, destruction of human lives, annihilation of plants and animals, and destruction of the environment."[63]

Similar criticisms came from individuals outside the official conference who were participating in the informal Environment Forum, where prominent environmental activists and nongovernmental organizations voiced the myriad environmental concerns of constituents not represented at the United Nations.[64] Also referred to as an "earth forum" and a "counter-conference," the event included almost five hundred accredited NGOs—many of them American, including the Hog Farm hippies—and scores of prominent individuals, including some official conference observers and delegates.[65] One jilted environmentalist excluded from the official U.S. delegation called it "a sort of environmental Woodstock for the world's affluent youth."[66] Participants espoused diverse objectives and viewpoints, but most opposed the war in Vietnam and supported binding limitations on nuclear weapons. They held a major protest of the Vietnam War in Stockholm during the second week of the conference.[67]

Perhaps more surprising than the invective hurled at the United States for its involvement in Vietnam was China's unexpectedly aggressive overall anti-Western position at the conference. Sino-American relations had thawed remarkably under Nixon, and the president had visited the communist nation in February of 1972. But Tang apparently hoped to use the conference to establish the People's Republic of China (PRC) as an important force in the United Nations, and he and his colleagues staked out a position as champions of the Third World in the face of "imperialistic superpowers." At the time of the conference's initial preparatory meetings in 1971, the People's Republic of China had not yet been admitted to the United Nations, and Tang insisted on reopening the all but finished "Declaration on the Human Environment" for "more democratic" revision now that the PRC had been seated.[68] Once reopened, negotiations over the language of the declaration became a forum for criticism and dissent, with the United States as a primary target. Despite the concessions of the Founex Report, many less developed countries still harbored deep reservations that a U.N. environmental effort would not sufficiently account for their development needs. China fueled their discontent. "We are firmly opposed to the superpowers subjecting other countries to their control and plunder on the pretext of improving the human environment," Tang declared.[69] Some African nations, Algeria most vocal among them, had begun to demand compensation for the historical exploitation of their environments by colonial powers, and Tang supported these demands.[70] "Victim countries," as Tang called them, "have the right to apply sanctions against and demand compensation from the culprit countries."[71] Because of the Chinese delegation's focus on the interests of less developed countries, its critical remarks drew heavy applause from an audience whose Third World representatives outnumbered the developed world by more than two to one.

The U.S. delegation handled these developments rather badly. Tang Ke's scathing remarks proved particularly damaging, not least because Russell Train and his colleagues had to clear unscripted statements with the State Department, the Department of Defense, and the White House before offering a rebuttal. The sequence of events that followed was embarrassing. The U.S. delegation asked for rebuttal time immediately after Tang finished speaking; but after nearly five hours of deliberation, Christian Herter of the State Department had to ask that the rebuttal be deferred for the weekend. As an alternative, Herter called a news conference for that afternoon, but he had to cancel at the last minute because the delegation had not come up with an acceptable message for the press. Train's rebuttal, finally delivered the following Monday, only led the press to revisit China's criticisms, and it contained little of substance. As it happened, the Chinese had already walked out of the conference, during a speech by delegates from what the Chinese called the "puppet clique" of South Vietnam, before Train had a chance to present his case.[72] A photo of diplomat and former child star Shirley Temple Black applauding Train showed row upon row of empty seats.[73]

THE LIMITS TO "GLOBAL"

It is easy to lament, as Russell Train did in his 1972 evaluation of the U.N. Conference on the Human Environment, the "politicization" of the global environment that occurred at Stockholm. In the years since, many scientists studying climate and the global environment have echoed Train's sentiment, complaining that political chicanery has continually infringed upon rational, science-based plans for environmental management and protection. But this is a misguided lament. Like the problem of climate change, the global environment as a concept has always been political, thanks in large part to the scientists who first helped to describe it. The SCEP and SMIC studies, after all, were undertaken with the express intent of informing a nascent political discussion about the global environmental crisis. *The Limits to Growth* was nothing if not politically prognosticative. Scientists' idea of a world made up of complex, interrelated, large-scale systems helped to buttress Maurice Strong's "new globalism" in international politics by highlighting mutual interest and the necessity of international cooperation in dealing with global problems. Scientists themselves explicitly framed their approach in such internationalist terms. It is unfair to impugn the actual scientific work involved in these reports as somehow biased or nonobjective, but the projects themselves arose self-consciously from a desire to put the global environment on the political agenda in the 1970s, and it would be naïve to ignore the political motivations of those involved.

For the scientific community, perhaps a more honest lament than Train's would be that the scientists who studied and described the global environment did not play a greater role in defining its politics. Along-side the U.N. leaders they supported, systems scientists found that their vision for cooperative scientific and political approaches to world-scale environmental problems conflicted with a geopolitical system that privileged national- and regional-scale economic concerns. The regime of global environmental governance established at Stockholm ultimately reflected the interests of these established constituencies at least as much as it did a collective concern for humanity's impact on the whole earth.

Then again, maybe this was not such a bad thing, especially in the short run. After all, a new framework for global environmental politics was established at Stockholm, and it was established with some success. As Anthony Lewis of the *New York Times* reported in his recap of the event, "One Confused Earth," the very occurrence of the conference—especially after the repeated confrontations between East and West, rich and poor, communist and capitalist in the preceding months—was a victory for international environmental consciousness. The event drew more than a hundred nations together to discuss a relatively new set of disconcerting global issues for the first time, and it achieved a good deal of real political success.[74] U.N. member states established benchmark regulations for ocean dumping, whaling, and toxic wastes, and the conference provided twenty-seven ambitious guiding principles for future international environmental agreements. The meeting also established the machinery for developing and implementing international environmental policy within the new U.N. Environment Programme (UNEP).[75]

The conference did little to lessen the dominance of nation-states' interests in international politics, but the concept of nonrenewable "world resources" began to resonate with delegates from both developing and developed nations. Growing Third World political interest helped to buoy First World environmental initiatives, both within UNEP and related U.N. agencies like the Food and Agriculture Organization and the World Health Organization. By incorporating developing nations and their concerns about the social and political ramifications of environmental degradation and protection into an international political process, the U.N. Conference on the Human Environment helped to establish the global environment as a significant geopolitical interest.

The scientific community may have been overshadowed at the conference by the development issue, but they did not exactly lose out at

Stockholm. Despite some nations' concerns over sovereignty and security, Earthwatch reinforced scientists' efforts to monitor and study the environment at a global scale, especially within the World Meteorological Organization. UNEP promised to help focus the process of international environmental policymaking such that policymakers might better access scientific information, and media coverage of the conference helped stir up public interest in international environmental issues involving the oceans and the atmosphere. It was hardly the New Atlantis that some had quietly dreamed of during the SCEP and SMIC meetings, but it was a start.

These stirrings of scientific success proved limited, however. As the 1970s wore on, atmospheric scientists found that the political framework for global environmental governance established at Stockholm, dominated as it was by national and regional interests, made it difficult to gain support within the United Nations for the truly global environmental problems involving the atmosphere—ozone depletion, aerosols, and CO_2-induced climate change.

At first, climate took a back seat to other issues because it was as yet a novel problem that could barely even be described, let alone incorporated into an international system of environmental protection. In 1972, there were significant divisions within the scientific community over the true causes, extent, and direction of climatic change. Robert White agreed that the World Meteorological Organization should study the problem, but despite the urgings of the SCEP, SMIC, and SCOPE reports, it was not a primary focus of the international environmental community. To the public, it was also profoundly unsexy. Scientists presented the world with a complex, forward-thinking image of global atmospheric change in the run-up to Stockholm. At Stockholm, Walter Sullivan wrote about people using garbage bags to dress up buses like whales.

Soon, though, as the potential impacts of rising atmospheric CO_2 became clearer, it also became increasingly clear that the international environmental community, for all its globalizing rhetoric, had no way of dealing with a truly global environmental problem like climate change. In Stockholm, scientists and political leaders framed everything from waste management to soil depletion to land-use change to water pollution as parts of a larger, global environmental crisis. But while these problems may have crossed borders and represented a cumulative or aggregate threat to the global environment, both their sources and impacts could be mapped onto real geographical spaces controlled by mayors, governors, dictators, presidents, parliaments, community leaders, and bureaucratic agencies of all stripes. Pitched as global, these geographically specific environmental problems all also fit, albeit sometimes awkwardly, into an international politics centered on the economically interested nation-state.

The global atmosphere, by contrast—fluid, dynamic, and borderless—had no such attachment to local or regional politics. Changes in climate or atmospheric composition might affect environments anywhere and everywhere, and in 1972 the interests at stake were still unclear. In a battle to define the global environment in terms of competing development and natural resource interests, a global-level phenomenon not rooted in any specific national or even regional space had no interested constituency. So while climate change certainly fit into scientists' vision for a cooperative global research effort administered by a revitalized United Nations, it remained a relative nonissue in international environmental politics until the 1980s. The U.N. Conference on the Human Environment in that sense was a failed attempt to include CO_2 in global governance. Only after another decade of effort would scientists succeed in making climatic and atmospheric change relevant to the regional- and national-scale interests at the heart of international politics. And by that time, the battle to define the global environment—reincarnated in the post–Cold War economic terms of "sustainable development"—had seen dramatic changes in the geopolitical landscape in which it was waged.

4

CLIMATE, THE ENVIRONMENT, AND SCIENTIFIC ACTIVISM

IN 1982, THE SIERRA CLUB'S EXECUTIVE DIRECTOR, MICHAEL McCLOSKEY, sent around "Criteria for International Campaigns" as a guide for developing international initiatives under the club's growing International Program. McCloskey was the head of an organization that had made its name through grassroots organizing, and his criteria were not all that surprising. First and foremost, McCloskey argued, an international environmental goal must be achievable "within a reasonable time frame." The goal must be "*clear and discrete*; not ill-defined." Viable goals must not "cover impossibly large situations" and must deal "with conditions for which a legal or regulatory solution is possible." McCloskey demanded that the issues the organization took up abroad fit neatly into an institutional forum—legal or regulatory— that the Sierra Club had the competence and capability to influence. Issues, he emphasized, should strike the club's members as clear threats to their "deeply felt values." The source of the threat must be something that could be pinpointed, with the larger problem put into a real-life context accessible to concerned members—that is, nothing "too exotic nor overly technical."[1] McCloskey wanted the Sierra Club's International Program to be an extension of the group's efforts at home—that is, tackling clear and discrete, easily accessible problems with definitive short- and midterm solutions achievable through familiar patterns of public advocacy, legislation, and litigation. The problem of climate change was none of these things.

Atmospheric scientists had cast rising CO_2 and other forms of atmospheric change in terms meant to be recognizable to American environmentalists during the supersonic transport debate, and they had framed increasing CO_2 as part of a global environmental crisis in the run-up to the U.N. Conference on the Human Environment. Both of these reinterpretations of CO_2 brought scientists into conversation with environmentalists, but in neither case did atmospheric scientists capture the attention or the growing political power of the broader environmental movement.

In part, climate scientists remained aloof from mainstream American environmental politics because climate scientists generally defined "the environment" differently than environmentalists did; and as McCloskey's criteria demonstrated, as late as 1982 there was still a big gap between the two groups. Environmentalists continued to focus largely on Americans' day-to-day quality of life. Proper waste management, clean air, ample local parks and open spaces, and access to the nation's pristine wilderness areas increasingly defined the "good life" in America, and environmentalists saw degradation from industrial pollution and irresponsible development as a threat to these "backyard" environmental amenities.[2] Climate scientists, on the other hand, addressed the environmental crisis primarily in terms of the food, water, and other natural resources essential to life itself. They sympathized with the broad, humanitarian development goals articulated by less developed countries at the 1972 Stockholm Conference. Just as they did internationally, domestically scientists hoped to use their expertise to help mitigate the detrimental effects of climatic variability on development and agriculture and to better plan for the potential impacts of long-term climatic change on human environments.

For their part, environmentalists initially responded tepidly to the issue of climate change. They hesitated to rally behind a loosely defined community of scientists whose definition of the environment differed from their own and whose specific political goals remained vague. Climate scientists often chose to work within the very agencies environmentalists sought to challenge, and they usually did so with few recognizably "environmental" objectives.

The nature of climate change itself exacerbated the problems of incorporating rising CO_2 into environmentalists' national and international political strategies in the 1970s. Members of groups like the Sierra Club and the Wilderness Society took an interest in international environmental issues; but like McCloskey, the leaders of America's environmental organizations selected their campaigns carefully. They looked for clear problems with definitive solutions. What Charles David Keeling had noted about the problems of understanding CO_2 as a pollutant in 1963 had not changed by 1980. Climate change was highly technical, global in scale, and rife with scientific uncertainties. It occurred over the course of decades and centuries. No obvious political or legal structures existed to implement solutions to the problems of climate—in fact, few if any solutions presented themselves at all. Until the mid-1980s, climate change—by then primarily a concern about global warming caused by rising CO_2—remained almost entirely within the purview of the scientific community.

Initially this community found remarkable success in using science itself to gain influence within the top levels of government and resources for further research. But success was fleeting. The early 1980s would reveal the profound political liabilities of the science-first approach to advocacy that characterized the politics of climate change.

THE DISCIPLINARY LANDSCAPE OF CLIMATE SCIENCE

To understand how science dominated climate change discourse in the 1970s and early 1980s, it is helpful to briefly explore the landscape of climate science itself as it developed during the period. In the early 1970s, making meaning out of the Keeling Curve was still a dynamic and controversial process within the scientific community. The causes, extent, and even direction of climate change—colder or warmer—were hardly certain. Scientists from different disciplines approached the study of climate with different and sometimes competing methodologies, and they often disagreed over what exactly they should study, how they should go about studying it, and what their resulting findings told them. With limited funding available for climate research, disciplinary and methodological divisions defined the climate science community of the early 1970s as much as did shared concerns over climate change and the human environment.

Perhaps the most well-known climatic controversy of the 1970s was the dustup over "global cooling"—a sideshow in the broader history of global warming but worth investigating for what it reveals about the development of climate science. Much has been made of reports from the first half of that decade that the most dangerous form of anthropogenic climate change might come, not from CO_2-induced warming, but from scenarios that could send the mercury the other way. Today, skeptics of global warming often point to this early uncertainty as evidence of the climate science community playing Chicken Little and claiming that the sky—or the temperature—is falling based on insufficient or uncertain evidence. These critics tend to use the decades-old story to suggest, despite a propensity of findings to the contrary in the years following, that even today there is considerable debate over the cooling theory. In fact, there is not. In rebuttal, some scientists and science journalists have gone out of their way to expose the "myth" of a cooling consensus, arguing that comparatively few scientists supported the 1970s cooling thesis and that the community as a whole never subscribed to the theory.[3] Their chief complaint—that global warming skeptics and deniers have willfully and invidiously misinterpreted this episode in the early history of climate science to serve their own purposes—does hold up, for the most part, under scrutiny. Even so, the story of the warming/cooling debate is more dynamic than either side lets on.

In the early 1970s, climate scientists generally agreed that climatic change could adversely affect the human environment, but their research produced two conflicting scenarios. The first, espoused primarily by atmospheric scientists working with advanced computer models, involved an overall *increase* in global temperatures caused by increases in atmospheric CO_2 and other greenhouse gases. The second, promoted largely by climatologists and geologists working with both models and physical and documentary evidence of past climatic shifts, described a relatively sudden *decrease* in global temperatures caused by potential combinations of volcanic dust, industrial pollutants, land-use changes, long-term cycles of solar radiation, and abrupt changes in the circulation of water in the oceans. Divisions between atmospheric scientists and historical climatologists were not hard and fast, however. Atmospheric models at different levels of sophistication also sometimes disagreed, alternately predicting warming or cooling. CO_2-induced warming remained the most likely scenario throughout the 1960s and 1970s, but as late as 1975 the scientific community had yet to definitively rule out the cooling problem.

Atmospheric scientists, whose models highlighted the warming effects of atmospheric CO_2, dominated the study of climate in the late 1960s and early 1970s. As early as 1967, Sukiro Manabe and Richard Weatherald, both at the Geophysical Fluid Dynamics Laboratory at Princeton, had used numerical models to confirm the long-standing theory that increases in atmospheric CO_2 would lead to increases the global mean temperature.[4] By the late 1960s, Charles David Keeling had collected a decade of annually oscillating CO_2 data that showed—as he, Roger Revelle, and Hans Suess had predicted—that CO_2 had risen significantly since 1958. According to the Manabe-Weatherald model, if the increase continued to the point of doubling, global temperature could rise approximately 2.3°C by the turn of the century, all other things being equal.

But all other things were not equal. Meteorologists' day-to-day temperature measurements revealed that although the early twentieth century had experienced a modest overall warming, in the years between 1940 and 1970 the earth had actually cooled. The twentieth century, climatologists recognized, had seen remarkably stable climatic conditions that starkly contrasted with the variability of the geological past. Some feared that the earth's string of interglacial good luck was finally coming to a close, possibly with help from its human inhabitants. At a conference of the American Association for the Advancement of Science in 1968, Reid Bryson of the University of Wisconsin explained that this seemingly anomalous cooling could be attributed to particles of smoke and dust in the atmosphere, coming not just from volcanoes—which many climatologists agreed could be agents of climatic change—but also from human activities, most notably the slash-and-burn agricultural practices of less developed countries in the tropics. Cyclical variances in solar radiation and increases in volcanic activity seemed to many to be the prime movers of this overall cooling, but Bryson and others warned that deforestation and other changes in ground cover were modifying the amount of sunlight that reached the earth's surface, increasing the albedo (reflectivity) of large portions of the earth's surface and thus inadvertently lending nature an insidious hand in the cooling process.[5]

Other studies supported this view. In 1971, Ishtiaque Rasool and Stephen Schneider of the National Center for Atmospheric Research put specific forms of pollution into the equation, with surprising results. Rasool and Schneider (the same Schneider associated with the *Study of Man's Impact on Climate*) incorporated aerosols from industry—sulfur dioxide (SO_2) the most important among them—into a simple atmospheric model, and they, like Bryson, found that the cooling effect of the aerosols more than canceled out the warming effects of CO_2. Combined with smoke and dust from agricultural activities and volcanoes, industrial pollution might not only overpower the warming effects of CO_2; it might create a cooling sufficient, in the right conditions, to trigger another ice age.[6]

Challenged by these alarming results, modelers at NASA's Goddard Institute for Space Studies (GISS), the National Center for Atmospheric Research (NCAR), and the Geophysical Fluid Dynamics Laboratory (GFDL) redoubled their efforts to create models that could handle complex variables in order to prove or disprove the new theory. They started with one-dimensional models that described vertical columns of air independent of the influences of seasonality and landscapes—models like the one Rasool and Schneider had used to demonstrate the potential climatic effects of industrial aerosols.[7] Working with an influx of observed and measured data from the Global Atmospheric Research Program, modelers pushed their programs to incorporate new and different aspects of the global climate system. With every improvement in resolution beyond the static, one-dimensional models of the 1960s, new ideas cropped up about what questions the model might help to answer. Atmospheric modelers like Rasool, Schneider, Sukiro Manabe (GFDL), Joseph Smagorinsky (GFDL), Warren Washington (NCAR), Akira Kasahara (NCAR), and James Hansen (GISS) worked to increase the capacity of their models to deal with large- and medium-scale externalities and feedbacks from throughout the system. With increasing computer power, they made remarkable gains.

These increasingly sophisticated computer models tended to reaffirm atmospheric scientists' initial conclusions that increasing CO_2, rather than aerosols, represented the biggest threat to global climatic stability. In the new models, airborne solids presented enormous complexity. They did not all act to cool the earth. As some climatologists already knew, atmospheric dust by itself both reflects and absorbs sunlight. It can also provide the nuclei for the formation of a reflective cloud layer, which in turn, depending on the type of clouds, might reduce the solar radiation reaching the earth's surface, producing a cooling effect.[8] But the new models showed that the height, duration, and geographical distribution of aerosols also made a difference in what effect they had. Some aerosols contributed to cooling, as Bryson, Schneider, and Rasool had originally predicted, but others actually contributed to warming.[9] Schneider himself pointed out that some calculations in his paper with Rasool did not stand up to more complex models.[10] As modelers incorporated dimensions like variable humidity, seasonality, and differences between tropospheric and stratospheric behavior, the increases in CO_2 attendant to fossil fuel consumption appeared as a more powerful force than aerosols in affecting the overall radiation budget of the globe.[11]

Bryson and other climatologists still were not convinced; but by the late 1970s, warming seemed to many scientists the most likely climatic problem. As oceanographer Wally Broecker of Columbia University's Lamont–Doherty Geological Observatory wrote in *Science* in 1975, "A strong case can be made that the present cooling trend will, within a decade or so, give way to a pronounced warming induced by carbon dioxide."[12] By the late 1970s, Broecker's terminology had caught on, and *global warming* began to enter both the scientific and popular lexicons.[13]

Still, divisions persisted within the climate science community, and these divisions tended to reflect differences over methodologies, which varied between disciplines. Steeped in a tradition of measurement and empiricism, more traditional meteorologists and climatologists like Bryson and former Weather Bureau director of climatology Helmut Landsberg distrusted the theoretical nature of general circulation models, which failed, in their minds, to account for the observed conditions of the climatic past. Historical meteorologists, geologists, and climatologists all worked with computer models, but they preferred physical, documentary, and statistical evidence of real past events, which gave them reference points for addressing the possible causes of a contemporary climatic shift.

General circulation modelers looking at CO_2, though they disagreed emphatically over how to structure models and what those models might reveal, as a group retorted that the traditional standard of empiricism had no place in theoretical constructions of future climates. The atmospheric conditions of those climates—twice the atmospheric CO_2 in this case—had no corollary in human history. Instead of relying on the historical-statistical methods of geologists and other climatologists, general circulation modelers set theoretically plausible boundary conditions and then ran numerical simulations of atmospheric phenomena based on the laws of fluid mechanics. These models described simplified interactions between components of the atmosphere, including the dust and aerosols that so concerned Bryson.

The disciplinary divisions that shaped the warming/cooling debate also tended loosely to follow institutional lines. University climatologists like Bryson, H. H. Lamb, Thomas Murray, and Robert Matthews promoted further research (i.e., greater funding) for historical and statistical climatological studies in order to compete with atmospheric modelers at well-funded government and quasi-governmental institutions like NCAR, NOAA's GFDL, and NASA's GISS.[14] These disciplinary and institutional categories were relatively flexible, however, and they represented only one aspect of the internal divisions of the climate science community.

THE CONTINUING BATTLE FOR "GOOD SCIENCE"

In the 1970s, the primary locus of climate change politics was climate science; and in the politics of climate science, it was not the direction of change that mattered most but its severity. Both the warming and cooling scenarios threatened the stability of environmental, agricultural, and economic systems by increasing "climatic variability," or the year-to-year fluctuations of local and regional weather patterns.[15] To return briefly to Thomas Friedman's appealing term, the issue was not warming or cooling but "weirding." Outspoken scientists like Bryson and Schneider emphasized the need to incorporate the risks of climatic instability into national and international government infrastructures, regardless of the direction of change. More conservative scientists, in contrast—some conservative politically, some only professionally—were wary of the crisis mentality that pervaded popular scientific literature in the 1970s and of the unspecified government action it implicitly supported. *Nature* editor John Maddox, for example, lamented "the Doomsday Syndrome" among environmental scientists more generally. He equated works like Barry Commoner's *The Closing Circle* and Paul Ehrlich's *The Population Bomb* with sidewalk "sandwich boards proclaiming 'The End of the World is at Hand.'"[16] With the supersonic transport debate and the controversial *Limits to Growth* study fresh in their minds, Maddox and others feared that overzealous and premature policy recommendations, absent a more definitive scientific consensus, might undermine climate scientists' authority as experts and undercut government support for their projects. The scientific disagreement about climate change, like the SST controversy, soon also became a debate about "good science" and its relationship to policy.

Nothing underscored the climate science community's ambivalence about popular advocacy more clearly than the response to Stephen Schneider's 1976 *Genesis Strategy*. Convinced that the risks of waiting to mitigate potential catastrophe far outweighed the rewards of "ill-founded certainty," Schneider challenged what he called "scientific conservatism," criticizing the scientific community's commitment to "consensus" on poorly understood issues with high stakes, as with climatic change. "Consensus," he contended, "is a poor way to do science."[17] The book's central metaphorical conceit—a comparison of climate scientists' warnings to Joseph's Old Testament plea to Pharaoh to prepare for a potential famine by storing food—reflected a conscious abandonment of the "traditional role of the scientist to advance knowledge quietly" in favor of a bolder activism that mixed science and policy.[18] Schneider called on "our global society to build into our means of survival sufficient flexibility and reserve capacity to hedge against the repeated climatic variations that have been so well documented in history."[19]

Following the example of environmental scientists and biologists like Barry Commoner and Schneider's friend Paul Ehrlich, Schneider used the platform of climate science to expound more generally upon the same *monde problèmatique* at the heart of more comprehensive studies like *The Limits to Growth*. Schneider consciously ventured "beyond the confines of [his] academic training."[20] "The dangers associated with climatic change," they wrote, "are merely a part of the entire world *problèmatique*, and to view them in isolation from the rest of the world predicament would be to repeat the mistakes of many narrowly specialized observers who have examined the prospects for the future *only through the tunnel of their expertise*."[21] Schneider used his position as a climate scientist to lend credibility to his analysis of the interactions between climate, food, energy, and politics. He then laced this analysis with pointed commentary and policy proscriptions that went far beyond the call for more funding that was traditionally acceptable within the scientific community.

With book jacket endorsements by Ehrlich, Margaret Mead, and Carl Sagan, *The Genesis Strategy* reached a remarkably wide audience. The book earned reviews in the *New York Times* and the *Washington Post* and landed Schneider, like Ehrlich and Sagan before him, in a seat across from Johnny Carson on *The Tonight Show*.[22] But the book also raised the ire of many atmospheric scientists, who were already up in arms about the proper relationship of scientists to society after the SST debate. Helmut Landsberg, director of the University of Maryland's Department of Atmospheric Science and Meteorology, disagreed with both Schneider's science and his popular approach. Landsberg was particularly critical of Schneider's "use of climatic models for predictions to guide public policy."[23] Quoting widely respected global circulation modeler Joseph Smagorinsky, Landsberg reminded Schneider that "crude or premature estimates can be very misleading in providing guidance or such far-reaching decisions and may be far more damaging than no estimate at all."[24] He emphasized Smagorinsky's warning that "we should be wary of basing broad national or international decisions on hand-waving arguments or back-of-the-envelope calculations."[25]

Conservative scientists like Landsberg attacked Schneider primarily for prematurely and recklessly venturing into policy and popular culture with highly uncertain scientific conclusions, but they also objected to his left-leaning policy proposals. In what Canadian climatologist F. Kenneth Hare called "devastatingly naïve" policy recommendations, Schneider advocated an array of progressive measures that would expand the size and power of the federal government.[26] Schneider semiseriously proposed a fourth branch of government, the "Truth and Consequences Branch," a scientific body charged with studying and providing information on the long-term impacts of current policies and actions. He also called for a government-funded U.S. grain reserve, various global environmental treaties, an international inventory and collective control of nuclear materials, and a more equitable distribution of development aid and technology between the First and Third Worlds. In short, he proposed a new national and international order based on science and reason, but one also implicitly infused with his own internationalist liberal values.

"This is a risky game for a younger man to play," cautioned Hare in his otherwise laudatory review of *The Genesis Strategy* in the *Bulletin of the American Meteorological Society*, "and he takes his professional life in his hands when he does so. For there is a widely held view that scientists ought to stay out of politics.... We are a conservative profession.... Schneider will not lack for criticism and even abuse for having written this book."[27]

As Hare and other sympathetic colleagues feared it would, Schneider's book drew criticism from higher-ups at NCAR who, according to Schneider, did not find his style appropriate for an NCAR research scientist.[28] His appearances on television, and particularly on *The Tonight Show*, further stoked resentment. Adversaries at NCAR later tried to deny Schneider's bid for promotion to senior scientist based, at least in part, on their distaste for his popular science. Schneider maintained his favored status among well-respected and progressively minded scientific leaders at NCAR, such as William Kellogg and Walter Orr Roberts; but as a colleague recalls, "even at NCAR, Steve was pretty isolated for a while because of that book."[29]

CLIMATE, FOOD, AND THE ENVIRONMENT

Internal disputes over "good science" and the limits of scientific advocacy did not play out in a vacuum. Behind the internal disputes that alienated Schneider lay a familiar anxiety about the relationship between climate science and environmental advocacy. In the 1970s, climate scientists' advocacy for more research on global climate change developed in tandem with an international environmental movement focused on global environmental degradation, and it was this international movement that provided the context for scientists' forays into environmental politics. In the late 1960s and early 1970s, scientists who studied elements of climate and weather sought to apply their basic research to a global environmental crisis that included practical problems of energy use, agricultural waste, industrial pollution, and ecological change. It was these types of practical issues—or a subset related specifically to climatic variability—that most concerned climate scientists in the 1970s.

At the time, the most important practical problem related to the climate and climatic change was food supply. Over the course of the 1960s, the world's population had continued to grow at an alarming rate, but technological developments in agriculture allowed the world to provide for these new inhabitants. The "Green Revolution"—the application of technological advancements in agricultural science to developing-world food production—alongside generally favorable weather, made feeding new mouths possible.[30] In the late 1960s and early 1970s, however, a series of climatic and meteorological anomalies led to alarming food shortages. In the African Sahel region, four consecutive years of drought produced collapses in the regional food supply. The shortages resulted in large-scale migrations, social unrest, and starvation. In 1972, the Indian monsoon came a week late, stressing a food infrastructure already operating on a thin margin. A drought in the Soviet Union also caused major food shortages, leading to mutually embarrassing purchases of eighteen million tons of grain from the United States, where domestic food prices began to rise. For a brief period in late 1973—just before the Christmas Eve oil price shock of that year—a bushel of wheat sold for twice the price of a barrel of oil.[31] Lester Brown of the Overseas Development Council estimated that in 1972, climatic variation reduced the world food supply by about 1 percent—the first such reduction in more than a decade.[32] "No single factor has a greater impact on food production in any country than weather," Brown wrote. "The vulnerability of the supply-demand balance to the weather suggests that the climate itself might well replace pollution as the dominant global environmental concern."[33]

The atmosphere may have received short shrift at the 1972 U.N. Convention on the Human Environment; but because of climate's relationship to food, climate change earned new significance as a geopolitical issue in the lean years of the late 1970s. The United States supplied three-quarters of the world's grain exports in that decade; and with U.S. grain stocks dwindling and the Department of Agriculture pressing acreage previously held out of production into service, foreign policy experts—including CIA analysts—worried about the international political consequences of another poor harvest.[34] "The stability of most nations is based upon a dependable source of food," read a CIA working paper, "but this stability will not be possible under the new climatic era."[35] The agency largely subscribed to Bryson's work on aerosols and by 1974 "had obtained sufficient evidence" of a global cooling trend to put climatic change on the geopolitical radar. "Climate is now a critical factor," the CIA confirmed. "The politics of food will become the central issue of every government."[36]

CLIMATE CHANGE AND AMERICAN ENVIRONMENTALISM

American environmentalists also took a keen interest in global environmental issues in the 1970s, but they responded slowly and cautiously to the issue of climate change. The U.N. Conference on the Human Environment revealed deep concerns within America's nongovernmental environmental organizations over global population, land use, species protection, ocean dumping, toxic wastes, and deforestation, but these concerns did not extend to climate change in a meaningful way. Throughout the 1970s and early

1980s, the Sierra Club, Friends of the Earth, the Natural Resources Defense Council, and the National Wildlife Federation all lobbied Congress vigorously to extend and expand the United States' ever-shrinking contributions to the U.N. Environment Programme, and many of these organizations created committees or sections to deal exclusively with international issues.[37] In general, however, American environmental groups continued to focus primarily on the local issues that interested their largely middle-class memberships, and they carefully selected the international issues they thought their members would support. Professional environmentalists sought tangible environmental problems with definitive solutions that appealed to the interests and values of their grassroots constituencies, and in the 1970s the issue of climate change did not yet fit the bill.

The Sierra Club's International Program was a central node of the American environmental effort abroad. "No organization did more to prepare the way for the Stockholm Conference and the creation of the United Nations Environment Programme than the Sierra Club," wrote UNEP chief Maurice Strong to Sierra Club International Program director Patricia Scharlin upon leaving his post in 1975, "and none have been more effective or consistent in their support for our international activities during the crucial formative period."[38] During the 1970s, Scharlin and her colleagues collaborated with the Natural Resource Defense Council, Friends of the Earth, Zero Population Growth, the National Wildlife Federation, and others to influence international environmental policy both within official U.S. government delegations and as an independent supranational bloc at a series of U.N. conferences on food, water, desertification, human settlements, and population.

The Sierra Club's commitment to population control helped drive its International Program. The club had helped to publish and promote Paul Ehrlich's *Population Bomb* in 1968, as well as follow-up articles by Ehrlich and his wife, Anne, in the early 1970s. When the group's International Program began to reevaluate its mission in 1976, members made it clear that population was their top international priority.[39] Tying overpopulation to overconsumption, many members saw population control—both in the developing and more developed countries—as the sine qua non of environmental protection. It was an issue of global proportions that underpinned environmental issues at all scales. "Most of our environmental problems are the *direct* result of over-population," one Sierra Club chapter officer wrote. "The root cause of most problems," another claimed. "This seems to me a basic issue which must be solved or all will be lost!"[40]

Only slightly behind population on the list of international priorities were concerns over what the Sierra Club called the global commons—those areas shared by many nations but technically owned by none, consisting mostly, in the Sierra Club's view, of ocean environments. Members worried that the pollution and degradation of the oceans and the loss of marine environments—along with air pollution and environmental degradation in Antarctica—might affect human and natural systems on a variety of scales. They encouraged the Sierra Club to defend these environments on the international stage.[41] In a 1976 survey of local-level Sierra Club officers, well over half of the respondents identified global commons issues as their primary international concerns, beating out geographically specific problems like deforestation and even nuclear proliferation. After population, concerns about the health of the oceans and air dominated the Sierra Club's International Program in the late 1970s, and so it would have made sense for the organization to become involved in the mounting national and international discussions about climate change.

But that did not happen. Few of the organization's chapter officials advocated a wholesale commitment to the international arena for any cause, let alone one as nebulous and uncertain as climate change. Roughly 20 percent of Sierra Club members polled in 1976 recommended that the organization commit 10 percent or more of its overall budget to international issues, but almost an equal percentage thought the club should reduce its international budget to less than 3 percent of the total, and just shy of a third hoped to keep international activities to less than 5 percent of the budget.[42] Though scientists and members of Congress became interested in the CO_2 problem early in the decade, the issue of climate change had not sufficiently piqued club members' concern in 1977 for the International Program to include climate as a major part of its next five-year plan. Nearly a third of the 1976 survey's respondents identified global commons issues as their top priority for the International Program, but the subheading "air pollution and climatological studies" drew no votes as a top priority, and less than a third of respondents identified climate as a priority at all.[43] As late as 1979, the year of the World Meteorological Organization's World Climate Conference, discussions of the climate and climatic change within the Sierra Club remained brief, tentative, and noncommittal.

The problem with making climate a Sierra Club issue, as Patricia Scharlin saw it, was that members did not see climate change as a "'back yard' kind of issue, or one that [commanded] a short term horizon."[44] Despite a strong rhetorical and administrative commitment to international challenges, the Sierra Club remained an essentially local grassroots institution. Scharlin noted that climate change "inspires emotional interest, has a global, international dimension, and very serious, long-term complications." But she also noted that the issue lacked the immediacy of problems like population growth or industrial air pollution that could easily be framed in terms of their local and regional impacts.[45] An increasing population, for example, not only strained the world's natural resources in the abstract; it also put actual pressure on those resources contained in America's wilderness areas and open spaces, threatening the wilderness experience itself. The problem of population growth was global, but its most important environmental impacts were local. As Scharlin pointed out, neither scientists nor environmentalists had yet managed to identify the specific threats that climate change might pose to local and regional environmental amenities.[46] As a result, Sierra Club members did not identify with the issue, and climate change failed to show up either in the International Program's goals or in larger club priorities.[47]

THE FORCING FUNCTION OF KNOWLEDGE

The scale and complexity that made climate change so difficult for environmentalists to engage with were the very qualities that appealed to many climate scientists. The uncertainties of climate change made scientists keen to study the issue and identify its potential impacts on natural resources. Within the American scientific community, efforts to incorporate climate science into resource management took two forms. First, most climate scientists, regardless of their politics, supported the proposed National Climate Program, a national-level bureaucracy for coordinating climatic research and incorporating climate data into domestic agricultural planning and natural resource management. The National Climate Program would not only support the needs of American farmers, ranchers, and emergency management professionals; it would also establish a secure role for climate scientists in formulating natural resource policies, both within federal agencies and in Congress. Second, scientists at the

American Association for the Advancement of Science hoped to extend these practical domestic applications of climate science to international development. The AAAS made climate change a central component of its international advocacy program, and the group worked to establish a firmer scientific consensus on CO_2-induced warming to buttress potential policy recommendations. For advocates of both the National Climate Program and the AAAS initiative, the immediate concern was to facilitate more and better research into climate. Their logic resonates like the chorus of a Greek drama. Better knowledge, climate scientists believed, would shape good policy.

In the mid-1970s, the National Academy of Sciences (NAS) released three reports on climate and resources that emphasized the practical applications of climate science. *Understanding Climatic Change: A Program for Action*, published in 1975, called for integration of government-sponsored climate research across disciplines and institutions in order to foster a better understanding of atmospheric phenomena and to provide useful data to policymakers and planners in areas like agriculture and emergency management.[48] The NAS Committee on Climate and Weather Fluctuations and Agricultural Production published *Climate and Food: Climatic Fluctuation and U.S. Agricultural Production* in 1976.[49] And in 1977, the NAS Panel on Water and Climate within the Geophysics Study Committee produced a study titled *Climate, Climatic Change, and the Water Supply*.[50] The reports contained technical and research recommendations, and all three advocated better coordination of climate research to facilitate its practical incorporation into planning at local, state, and national levels.

The NAS studies were in part prompted by members of Congress who supported coordinating climate research in the service of natural resource planning. Congressional Democrats saw the increasingly frequent and dire—and often contradictory—reports and predictions from climate scientists, along with the lack of national coordination in climatic research, as an opportunity to take the initiative on federal science policy and undercut executive control over government science agencies.[51] Led in the House by California representative George Brown, concerned congresspeople noted that "the present research and data-gathering in the federal government is scattered, fragmentary, and inadequate compared to the impact of ignorance in this area."[52] Brown and his colleagues in the Senate pushed for a congressionally legislated National Climate Program that would establish the framework for national-level coordination of climatic research, monitoring, and warning systems.[53] In 1976, Brown and fellow House Democrat Philip Hayes introduced the National Climate Program Act with bipartisan support. The act's promoters accepted that "longer term changes in climate, whether occurring naturally or resulting from human activities, or both, may be leading to new global climate regimes with widespread effects on food production, energy consumption, and water resources."[54] They recommended that the president appoint an agency—be it NOAA, the NSF, NASA, or whomever—to oversee government-sponsored research pertaining to climatic change and the problems it might induce.[55]

Much of the congressional interest in the National Climate Program Act came, not from believers of the "prophets of doom," but from representatives of largely rural states who wanted scientific agencies to meet their constituents' practical everyday needs. Members of Congress like Brown and Republican senator Harrison Schmitt of New Mexico expressed grave concerns over deforestation, desertification, ice cap depletion, and other large-scale environmental problems associated with climatic change, but most politicians focused on the more mundane aspects of climate that affected their states' farmers and ranchers. They wanted to know about extremes of heat and cold, about storms and severe weather events, and, more than anything else, about water. Climate scientists and congressional leaders alike saw an immediate need for creating and distributing climatic data to "users" in short- and medium-term agricultural and hydrological planning. As Ed Epstein of NOAA's Environmental Monitoring and Prediction Unit—and later head of the National Climate Office under the National Climate Program Act—noted, there was certainly a demand. Requests for climatic data from the extant National Climate Center had climbed from 2,500 per month in 1972 to 5,700 per month during the first quarter of 1977 alone.[56]

In the late 1970s, the AAAS decided to make climate change the focus of a long-term international initiative in resource management and environmental protection. The largest general scientific society in the world and the publisher of the journal *Science*, the AAAS was a member-driven advocacy organization committed to advancing science in the public interest. Unlike the National Academy of Sciences, which responded mostly to the scientific needs of Congress and other parts of the federal government, the AAAS tried to reflect the scientific interests of a larger American public.[57] In the late 1960s and early 1970s, the AAAS had taken a turn to the left, passing resolutions opposing the Vietnam War and supporting environmental protection, civil and human rights, and gender equality in the sciences. For the most part, however, the organization focused on less volatile practical programs for enhancing international scientific cooperation, improving domestic science education, and increasing the public's overall engagement with science and technology. As the AAAS rebounded from the tumultuous 1960s, it established a Committee on Future Directions in order to refocus on these core scientific, political, and educational activities.[58] After relatively successful but ad hoc AAAS participation in U.N. meetings on population, human settlements, and desertification following the U.N. Conference on the Human Environment, the committee agreed to make CO_2-induced climate change a programmatic focus.

The committee's reasoning inverted that of the Sierra Club. The multidisciplinary subject of climate represented the sort of "long-range scientific effort of international consequence" that could "command the interest of a wide segment of the association's membership." As a result, the AAAS established a Committee on Climate as its premier international initiative, naming former AAAS president Roger Revelle as chairman.[59]

If there was an institutional embodiment of science-first climate change advocacy, it was the AAAS of the 1970s. Revelle and his colleagues at the AAAS envisioned the organization as a nexus for American scientific, political, and practical interest in the climate and climatic change, with more and better research at its core.[60] In 1978, the group identified no fewer than ten federal agencies involved in climatic research: the Departments of Agriculture, Commerce, Defense, Energy, Transportation, the Interior, and State; the Environmental Protection Agency; NASA; and the National Science Foundation. Under the auspices of the National Climate Program, these agencies had begun to coordinate their research with each other and with state agencies and scientific institutions, but they had yet to develop a mechanism for dealing with the various international groups—mostly U.N. agencies—interested in climate change. These groups included the World Meteorological Organization, the World Health Organization, the U.N. Environment Programme, and the Food and Agriculture Organization. The AAAS saw gaps between earlier mission-oriented government projects like the Climate Impacts Assessment Program, a limited and often overly academic NAS initiative, and the complex bureaucracy of the international research effort at organizations like UNEP and the WMO. Revelle and his advisory group—including prominent members of the climate science community such as former NOAA chief Robert White, NCAR's William Kellogg, Eugene Bierly of the NSF's Climate Dynamics Research Section, and Congressman Brown and his staff director and science advisor, Tom Moss—proposed that the AAAS serve as a link between these various organizations. They hoped the AAAS could build, in White's words, "the bridge between the complexities of the science and the general public and the policy-makers."[61]

The AAAS committee emphasized the importance of international cooperation and interdisciplinary communication. "It would be hard to find a problem in science that has such global characteristics," White contended.[62] Data collection and analysis required input from scientists studying disparate subjects across the globe, and subsequent discussions about the policy implications of climatic research would necessarily include a diverse community of nations, including many from the developing world.

Revelle's advisory group went out of its way not only to frame climate in terms of resources but also to define climate science itself as a potential resource for less developed countries.[63] The AAAS saw climate science as an environmental management tool, much like the sciences of forestry or agronomy. "Climatology is a peculiarly applied science in that its raison d'être is its relation to human welfare," the group noted. "You have to find out those things about climate that mean something to somebody."[64] The committee believed that disseminating climatic information would help developing countries to better exploit favorable conditions when they existed and to protect themselves against anomalies when they occurred. Managing climate and climatic information required participation not only from governments and organizations but also from individuals responsible for managing and utilizing resources at the local level. With a U.N.-sponsored World Climate Conference upcoming in 1979, the AAAS hoped to establish better channels of communication between the World Meteorological Organization and the farmers and water resource managers most affected by the potential impacts of climatic variability.[65]

White and Revelle understood that political solutions to the global problems associated with climate change would require buy-in from the developing world, and this in turn would require an unusual degree of scientific certainty. Without discussing specific mitigation policies, the

AAAS climate committee noted that preventing climatic change or mitigating its impacts might include potentially disruptive restrictions on behavior and would require heavy capital investment. Committee members saw a need for "an unusually high degree of unanimity and clarity on the part of the scientific community to obtain from political leaders the decisions and actions required." And that unanimity needed to come not only from the atmospheric and geophysical scientists studying the climate but from the social and biological scientists addressing the possible "higher-order effects" of climatic variability on biological, social, and economic systems. The scientific community, argued Tom Moss, should strive for international cooperation in order to create a global scientific consensus on climatic change, because "unless there is a world science policy base for that kind of change, nothing is going to happen."[66]

Moss and others on the AAAS climate committee believed that a strong scientific consensus, broadly publicized, would lead to positive political action on climate change. This belief had driven climate research since Roger Revelle began lobbying Congress for research funding in the 1950s, and it would shape scientists' advocacy on climate change into the twenty-first century. "Political problems," as Moss put it, "will be moved by the 'forcing function' of knowledge. A worldwide *scientific* consensus of the possible impact of climatic trends will be the forcing function for the difficult *political* decisions, such as regulating land use."[67]

Through the "forcing function of knowledge," scientific consensus, presented in the appropriate governmental and institutional channels, would become political consensus. By helping foster better international cooperation and coordination of climate science, then, the AAAS would also eventually help make better climate policy.

THE VIEW FROM 1979

The AAAS advisory group was sanguine about the efficacy of scientific consensus in part because they saw such agreement about the basic climatic effects of CO_2 already forming within the American climate science community. Reports of the NAS Climate Research Board (CRB) linked CO_2 to warming throughout the mid-1970s, and by 1976 only a few scientists still held to the idea of global cooling. As one member of the CRB put it in 1979, "A plethora of studies from diverse sources indicates a consensus that climate changes will result from man's combustion of fossil fuels and changes in land use."[68] In *The Genesis Strategy*, Stephen Schneider had criticized the plodding process of consensus building as a poor way to do science, but Tom Moss and his AAAS colleagues now saw an opportunity to capitalize on the power of a unified scientific voice. They hoped to encourage consensus by bringing scientists from different disciplines together to hash out the presumably minor points on which they disagreed.

In hindsight, the AAAS group's faith in the power and possibility of consensus seems problematic at best. As scholars Naomi Oreskes, Erik Conway, and Matthew Shindell explain in their work on what they call the "social deconstruction of scientific knowledge," scientific consensus, for all its power, can become extremely vulnerable in the face of dissent.[69] Scientists rarely define the criteria for establishing consensus on an issue; but as was the case with the AAAS, unanimity is often the ideal. As long as an individual scientist can demonstrate her adherence to the standards of "good science"—political neutrality, objectivity, and a disciplinarily defined methodological rigor—her single dissenting voice can undermine the ideal of unanimity and cast doubt upon a scientific community's "consensus" view. The more powerful or prestigious the dissenting scientist, the greater the challenge to consensus. As both environmentalists and antitobacco advocates learned during the 1970s and 1980s, destroying consensus by manufacturing doubt is far easier than forging even an overwhelming majority of agreement, let alone a unanimous viewpoint.[70] In fact, much of the debate about global warming in the years since 1979 has been a contest over the meaning and value of scientific consensus.

For climate scientists, however, these were mostly lessons of the 1980s. In 1978, their bitterest battles over consensus still lay ahead. With a National Climate Program supported by both parties in Congress, a Democratic president in favor of renewable energy, and an increasingly interested world public, the AAAS saw an opportunity to turn good climate science into good climate policy by bringing together the diverse group of biological, physical, environmental, and social scientists working on problems of climate change.

The AAAS found an ally in President Carter's new Department of Energy (DOE). An exceptionally cold winter and a shortage of natural gas in 1976–77, along with increasing gas prices and inflation, confronted Carter at his inauguration, and his single term would be marred by energy-related crises.[71] Though his policies largely failed to protect the American economy from oil price shocks in the short term, Carter's newly formed DOE began to tackle significant environmental issues in its energy research.[72] The Keeling Curve soon began to show up in the nation's energy portfolio.

In 1977, the DOE's Office of Health and Environmental Research began a collaborative study with the AAAS on the climatic effects of an increase in atmospheric CO_2. This study of CO_2—actually a series of interdisciplinary conferences beginning with a workshop in Annapolis, Maryland, in the spring of 1979—provided the AAAS Committee on Climate with a national-level example of the kind of conference they hoped to sponsor at an international level. The effort had ample funding, widespread participation from the scientific community, and the promise of a direct conduit to policymaking within an important executive agency.

If there was a tinge of regret in program director David Slade's early assessment of the DOE-AAAS study, it was that the climate issue failed to attract many representatives from the environmental community. To many environmentalists, the DOE—the agency that managed the nation's nuclear energy and had a hand in many of its dams and coal-fired power plants—was the enemy.[73] Slade invited environmental groups, but they did not come.[74] With the exception of Thomas Kimball of the National Wildlife Federation, who officially supported the National Climate Program Act in a letter to Congress, environmental groups ranked the study of climate change a low priority and remained aloof from the climate debate into the early 1980s.[75] Even so, from where Slade sat, the prospect of turning climate science into responsible energy policy under Jimmy Carter looked pretty good.

TOP-DOWN, SCIENCE-FIRST

Environmentalists' tepid initial response to the issue of climate change underscored larger philosophical differences between professional environmentalists and politically active scientists over both the tactics and priorities of environmental protection. In the 1970s, American environmentalists thought primarily in terms of "protection," and their institutions reflected this defensive mind-set. Beginning in the early 1970s with the Sierra Club Legal Defense Fund and the Natural Resources Defense

Council, environmental organizations began to focus on congressional lobbying and legal action in response to immediate affronts to local and regional environments. Their strategy was one of reaction. Laboring diligently to influence new legislation and uphold the standards of existing laws in order to counteract corporate and governmental environmental transgressions, environmentalists planned future strategies of environmental protection based on existing or immediate environmental hazards. Because of their uncertainties, rising atmospheric CO_2 and ozone depletion—what Michael McCloskey referred to as "first-generation issues"— had not yet germinated into tangible threats, and therefore, garnered little in the way of strategic planning or fund-raising.[76]

Meanwhile, concerned climate scientists worked within rather than against executive bureaucracies. They were generally unburdened by the immediate needs of public constituencies, and they sought to illuminate the potential risks of what NCAR social scientist Mickey Glantz referred to as "those low-grade, but continually increasing, insults to the environment for which pluralistic society (the national as well as international community) has not yet found an effective policy-making process."[77] Rather than reacting to the impacts of climatic change, many of these scientists believed, governments and societies should work to preempt environmental catastrophe through mitigation and adaptation. Theirs was a gospel of preparedness.[78] Through the National Climate Program Act, the National Academy of Sciences, and the joint DOE-AAAS workshops, climate scientists worked to incorporate climate change into government policy from the inside out.

Scientists and environmentalists also differed over the relative priorities of economic development and environmental protection. Echoing the early twentieth-century debate between "conservationists" like Theodore Roosevelt and his Forest Service chief Gifford Pinchot and "preservationists" like Sierra Club icon John Muir, environmental leaders of the 1970s, including McCloskey, David Brower, and Thomas Kimball, saw a fundamental tension between rapid economic development and the health of the natural environment.[79] "While I think we should be very sensitive and understanding with respect to the need for economic advancement by the poverty-stricken," wrote McCloskey shortly after the 1972 U.N. Conference on the Human Environment, "I think that we are not in a position to be emphatic in stating that the two goals are entirely compatible. On the contrary, I think we see some very clear problems with respect to reconciliation, though I think we can express every hope that we can find a way to have both a tolerable environment and sufficient human economic security."[80]

Many of the scientists most vocal in debates about the impacts of climate change, meanwhile, shared a neoprogressive commitment to socially and environmentally responsible resource management that, in their eyes, could reconcile these tensions in the service of improving the quality of human life.[81] Despite the dire warnings embedded in their conclusions, studies by Roger Revelle, Mickey Glantz, Stephen Schneider, and others rested on the optimistic premise that with good scientific research behind them, policymakers could—and should—encourage new types of development in order to protect the coupled human and natural systems potentially affected by irresponsible—or "unscientific"—growth.

In retrospect, scientists' faith in the "forcing function of knowledge" to inspire progressive climate policy seems at best misguided, at worst hopelessly naïve. But at the time, it actually seemed to work. During the 1970s, climate scientists gained remarkable access to federal agencies and policy elites, and by the end of the decade they had begun to build the type of consensus that Revelle and his AAAS colleagues had hoped for. In 1979, a National Academy of Sciences committee headed by meteorologist Jule Charney of MIT released a comprehensive review of CO_2 and climate research conducted over the previous decade. The study concluded that a doubling of atmospheric CO_2 would lead to a 1.5–4°C warming of the earth, and that if atmospheric CO_2 continued to increase, there was "no reason to believe that these changes will be negligible."[82] The joint DOE-AAAS venture convinced David Slade, also as early as 1979, that the Department of Energy not only should continue to fund research on CO_2-induced climate change but also should develop a national plan for ameliorating and adapting to the unintended climatic consequences of fossil fuel energy consumption.[83] Slade laid out a potential research and assessment program that would support a continuing dialogue between climate scientists and policymakers through the end of Carter's theoretical second term. In 1979, *Nature* called climate change "the most important environmental issue in the world today," and Slade sought to establish a prominent place for the issue in America's ongoing discussion about its most pressing political and economic problem: energy.[84] Working within government, climate scientists appeared to be on their way to creating meaningful policies on CO_2 and climate change for the 1980s.

But climate scientists' commitment to working within government bureaucracies left them vulnerable to political change. Slade's program, like Carter's second term, was not to be. Despite strong talk of a responsible renewable energy program, Carter's troubles in the Middle East led him to back CO_2-producing synthetic fuels made primarily from coal. Carter's commitment to such fuels put Slade in an awkward position vis-à-vis presidential energy policy. When Ronald Reagan was inaugurated in January 1981, Slade's influence within the DOE decreased even further. Amid the new president's severe budget cuts to government-funded social science and environmental research, the director of the DOE's Division of Carbon Dioxide and Climate Research found himself isolated within an agency whose mandate under Reagan would be "cheap energy now!" "I am not paranoid," Slade quipped to David Burns in the fall of 1980, "but, besides wrapping my head in silver foil before retiring so as to ward off the rays the CIA beams at me nightly, I carefully review publications sent to me to see how 'they' are cutting off my information sources.... As the high priest of CO_2 for the federal government I view it as more than just a coincidence that my copy of science [sic] is missing the article on CO_2."[85]

5

||||||||||||||||||||||||||||||||||||||||||||||||||||||||||||

THE POLITICS OF DISSENT

THE ANNUAL OSCILLATIONS OF THE KEELING CURVE CAPTURE ONE OF
the more dramatic cycles of decay and renewal on earth. The curve is like
a negative image of the earth's plant life. At the annual peak of global
atmospheric CO_2, typically in May, the deciduous plants of the Northern
Hemisphere have only just begun to bud, stirred by warmth and sunlight
from a long winter dormancy. And then, almost suddenly, they explode
into summer growth. CO_2 hovers at its peak until early June, and then it
plummets steeply through the annual mean and into its late October or
early November trough, when the Northern Hemisphere's growing season
finally ends, and the expunged leaves and stems begin to return their CO_2
to the atmosphere.

The cycles of history rarely unfold so regularly, but in 1980 the fortunes
of scientists interested in CO_2 nearly matched the cycles of CO_2 itself. In
May of 1980, atmospheric CO_2 stood at around 341.5 ppm; and a year after
the first DOE-AAAS conference, its concentration had become an important concern for the Department of Energy. By November, however, CO_2
bottomed out at 335 ppm, and the American public had elected a president
for whom neither CO_2 nor other, more mainstream environmental causes
held much appeal. While CO_2 itself would rebound the following spring,
the fortunes of those studying the substance would rebound much more
slowly. When climate scientists' situation did finally improve, they found
that they had tied their fate to that of America's environmentalists and to
the Democratic Party like never before.

The partnership that developed between climate scientists and American environmentalists in the 1980s arose out of a familiar story unfolding in a new context. America's environmental groups took on the issue of global warming slowly and haltingly in the early 1980s, but by 1985 most prominent environmental organizations had begun to recognize climate change as a potential priority for the future. Shared scientific and environmental interests continued to bring climate scientists and American environmentalists into conversation in the 1980s, as they had since the early 1960s when Keeling framed CO_2 as a form of pollution.[1]

But in the new decade these two groups also came together to face a common political enemy. His name was Ronald Reagan. Elected in 1980, the new president harbored disdain for what he saw as an unnecessarily alarmist and antibusiness environmental movement. He had similar contempt for social and environmental science research. Upon inauguration, he sought to dismantle both the Council on Environmental Quality and the Department of Energy, and he replaced capable administrators at the Environmental Protection Agency and the Department of the Interior with zealous acolytes hostile to environmental regulation.[2] The administration aggressively downsized federal environmental research, especially targeting renewable energy. Unable to fully excise the DOE from the executive branch, Reagan's Office of Management and Budget reallocated more than a billion dollars of the DOE's money to the Department of Defense.[3] The ham-fisted approach left the DOE with neither the funding nor the personnel to continue its research on CO_2.[4] Like environmental scientists of every stripe, scientists involved in the joint DOE-AAAS climate study begun under Carter—the largest and most extensive program of climate research up to that point—saw their funding slashed and their reliable government contacts replaced by party-line political appointees. Having worked hard to incorporate climate change into the research and policy agendas of federal agencies since the 1950s, climate scientists suddenly and unexpectedly found themselves on the outs. As a result, in the 1980s the politics of global warming became, like environmental politics more broadly, a politics of dissent.

THE TRANSITION

Few individuals' stories capture the rapid and traumatic change that accompanied the transition from Carter to Reagan more completely than the experiences of Denis Hayes at the DOE's Solar Energy Research Institute (SERI). A jack-of-all-trades within the environmental movement, Hayes had been a key organizer of Senator Gaylord Nelson's 1970 Earth Day initiative. In 1977, a strange twist of circumstances brought Hayes into the orbit of the Carter administration. That year, Hayes published a book on renewable energy called *Rays of Hope: The Transition to a Post-Petroleum World*. It was well received, and while promoting the book Hayes met briefly with Saudi Arabia's powerful oil minister, Zaki Yamani, to whom he gave a signed copy. Hayes did not give the gift a second thought. A short while later, however, Jimmy Carter's Republican secretary of energy James Schlesinger found himself in a waiting room outside Yamani's office in a plush Saudi palace. There on the table was Hayes's *Rays of Hope*, the only reading material in English. A bored Schlesinger picked it up and read enough of it for Hayes's name to stick with him.[5] With boosters of nuclear energy at the head of the DOE nuclear program and men passionate about coal leading the fossil fuels program, Schlesinger was intent on putting someone passionate about renewables at the head of the DOE renewables program. When SERI needed a new director shortly after opening its doors in 1977, Schlesinger called Hayes.[6]

Hayes showed up to SERI ready to put renewables on the national energy map. As one of the his first tasks, he put together a study to determine how energy conservation and renewable energy production might affect the nation's overall energy use. Authored in large part by Hayes's assistant director, Henry Kelly, the study concluded that by the year 2000, an aggressive renewable energy policy could account for more than a quarter of the total energy production in the United States, and that the Carter administration should commit to a goal of producing 20 percent of all U.S. energy through sources like solar, wind, and geothermal power.[7] In short, SERI showed the DOE that renewables were a feasible option for the future.

The institute delivered the report—known to its authors as the Solar Conservation Report or the Sawhill/SERI report (named for Deputy Secretary of Energy John Sawhill)—to the DOE in November of 1980. November, however, was not an ideal time to deliver the report. In the wake of the 1980 election, the DOE tabled the study, and its conclusions were at least temporarily forgotten.[8]

To Hayes, the election at first looked like little more than a bump in the road for renewables. Like many environmentalists, Hayes had moderate hopes for the incoming Reagan administration—or at least he "wasn't sure things would be getting a whole lot worse."[9] Carter had backed off considerably on environmental policy in the waning years of his term, nowhere more than in his energy policy. In August of 1979, he had replaced Schlesinger with Charles Duncan, a Texas millionaire and member of the Coca-Cola Company's board of directors who, in Hayes's view, "had essentially no knowledge about renewables—or anything else related to the environment for that matter."[10] In a series of short radio addresses during the campaign, Reagan had specifically lauded renewables as a way to support the libertarian desires of "the great American loner"; and up until Reagan started picking his cabinet, many professional environmentalists thought they might be able to work with the new president. In any case, SERI and the DOE had four years of solid growth propelling them forward. Even without a supportive president, Hayes and his colleagues felt secure in their posts.

They were sorely mistaken. Upon taking office, the new administration slashed SERI's budget by nearly 80 percent and issued pink slips with two weeks' notice to scientists who had recently left tenured university positions to take their SERI posts.[11] The president began replacing officials at both SERI and the DOE with loyal political appointees. Most egregious was James Edwards, a dentist and former South Carolina governor whom Reagan tapped as secretary of energy to appease Republicans from the traditionally Democratic South who were upset that the president had no southerners in his cabinet.[12] In 1986, the president would add insult to injury by stripping the solar panels from the White House roof. For Hayes and his SERI colleagues, by as early as January of 1981 the DOE was beginning to look like part of the energy problem rather than a potential part of the solution.

Sitting on the as yet unpublished Solar Conservation Report, Hayes and Kelly decided to fight back. What Hayes recalled next reads like something out of a Raymond Chandler novel:

> So I'm back in Washington, D.C., early in 1981 for a meeting with this acting assistant secretary in the Department of Energy, a guy named Frank DeGeorge. I always go fishing for intelligence, so I went over to the department an hour early and I was kind of walking around and a

friend of mine—one of the deputy assistant secretaries—stopped me at his desk.

And he says, "Hey, have you talked to DeGeorge yet?"

And I say, "No, why?"

And he says, "The Solar Conservation Report? Yeah, it's dead as a doornail, they're never going to let that out."

So I strolled down the hall, nipped into the nearest elevator, went back to my hotel, called Henry Kelly and told him to Xerox as many copies as he possibly could, but get at least fifty or a hundred of these things FedExed out to environmental organizations and essentially anyone who might be interested. I had my secretary call this Frank DeGeorge's office and say I was sorry, I was going to have to cancel that meeting because I'd come down with the flu, but we'd like to reschedule it for the next time I'm in town. Didn't seem to be anything terribly important; he hadn't told me it was.

So I flew back to Colorado, and lo and behold I get a call from Frank DeGeorge, and we're talking about the report, and he says it's time to pull the plug on it. I said, "Geez, Frank, there's no way to do that. I've got a hundred of these things out there for review right now."[13]

In the short run, Hayes's ploy worked. He had seen to it that the report was in the hands of several environmental groups and key politicians, its intended audience. He was especially sure to leak a copy to Representative Richard Ottinger of New York, a member of the House Committee on Energy and Commerce and a strong supporter of federal conservation and solar programs. Ottinger made the report public domain by reading it into the congressional record. Brickhouse Publishing then released the now public document as *A New Prosperity: Building a Sustainable Future* in April of 1981.[14]

Reagan's staff was not amused. When Hayes told me the story in 2009, he explained the fallout with a wry smile. Hayes, it turns out, was the same Denis Hayes who had as a college student at Stanford in 1969 led the occupation of the applied electronics building, much to the chagrin of then governor Ronald Reagan, who had put the National Guard on standby in Redwood City. According to Hayes, when Reagan's White House counsel and former gubernatorial campaign manager, Edwin Meese, connected the name of the wayward SERI director with the brash student body president, he flew off the handle. Meese reportedly had a three-by-five card with Hayes's name and picture printed on it, and he carried it in his front pocket until Hayes was finally fired—or was asked to resign—for "not being a team player" in the summer of 1981.[15]

THE "REAGAN ANTIENVIRONMENTAL REVOLUTION"

Well-orchestrated retaliation aside, Hayes's fall from government grace typified scientists' and environmentalists' experiences throughout the federal government in the early 1980s. For climate science, the most important changes occurred in the Department of Energy, where the appointments radiated outward from James Edwards. Initially, the Reagan administration explored dismantling the DOE altogether, eliminating a wide variety of its programs and rolling what would be left of its core, an agency called the Energy Research and Technology Administration, into either Commerce or Interior.[16] Reagan's aides eventually preserved the DOE as a single cabinet-level entity, but the administration cut its budget by more than a billion dollars.[17] The administration's DOE budget request for fiscal year 1983 was an 87 percent reduction from what Carter had requested for fiscal year 1981.[18] These budget cuts severely curtailed the research of the DOE's Office of Health and Environmental Research, including David Slade's joint DOE-AAAS climate research program.[19]

By itself, the DOE-AAAS project was small potatoes, but the broader DOE funding cuts made waves throughout the federal research structure. DOE operated a number of national laboratories—Lawrence Livermore, SERI, and Oak Ridge, for example—and researchers at these laboratories scrambled to replace DOE money with funding from the Department of Defense ($3.8 billion richer in R&D dollars under Reagan), the National Science Foundation, and the private sector. For climate research in particular, this put exceptional strain on the NSF, which already funded the quasi-governmental National Center for Atmospheric Research and a number of university-based climate projects. Worse still for climate scientists, the Reagan administration ultimately controlled most of these resources too, and what little new money the NSF could make available for climate research began to go to projects investigating the direct impacts of rising atmospheric CO_2 on crops and managed forests—studies that in isolation made CO_2 buildup look a little less threatening and possibly even beneficial.[20]

The Council on Environmental Quality endured similar cuts in funding and changes in personnel, including a 50 percent reduction in staff and a budget slashed by two-thirds.[21] The deepest cuts came in the CEQ's research budget, a tacit rebuke to the kind of work the agency had undertaken under Carter in its comprehensive 1980 systems-based *Global 2000 Report to the President*. Not coincidentally, that report dealt in part with the potential environmental impacts of climate change.[22]

The Carter to Reagan transition had a profound impact on the DOE and CEQ, but the most notorious bureaucratic changes were initiated in what historian Sam Hays refers to as the "Reagan Antienvironmental Revolution," carried out in those agencies most responsible for the everyday management of the environment.[23] During the 1980 campaign, Reagan had mobilized moderate Republican conservationists like Russell Train and William Ruckelshaus—both former EPA administrators—to help him establish a positive environmental image. Once elected, though, he brought in highly conservative, pro-business, pro-industry advisors tied to the Washington, D.C.–based Heritage Foundation to reshape American environmental policy. The group sought to control policy by appointing loyal right-wing acolytes as policymakers. Reagan tapped Ann Gorsuch, a leader of the Colorado legislature's Republican Right and a lawyer for mining and agriculture interests, to head the EPA. Gorsuch, in turn, named fifteen like-minded subordinates, eleven of whom had ties to the very industries the EPA was supposed to regulate.[24] John Crowell, attorney for the world's largest buyer of national forest timber, took charge of the U.S. Forest Service. And Robert Harris, the leader of a legal effort to overturn the Surface Mining Control and Reclamation Act of 1977, became head of the Office of Surface Mining, the very agency he had been fighting.[25]

The antienvironmental revolution's coup de grace came in the person of Reagan's new secretary of the interior, James G. Watt. An evangelical Christian, founder of the quasi-libertarian Mountain States Legal Foundation, and dubbed the "Robespierre of Western Resistance" by conservative columnist George Will, Watt openly opposed American conservation efforts, both as a matter of policy and a matter of faith.[26] He described environmentalism as "a left-wing cult dedicated to bringing down the type of government I believe in." "My responsibility," he told the *Wall Street Journal* in 1981, "is to follow the Scriptures which call upon us to occupy the land until Jesus returns."[27] Watt committed the Department of the Interior to "mine more, drill more, cut more timber," and from his first day in office, he made it his mission to relax environmental regulations, promote natural resource exploitation, and reduce federal landownership to the greatest extent possible. As incoming secretary of the interior, he vowed to "strike a balance" between the interests of environmentalists, recreationalists, business, and industry, but for environmentalists like Russell Peterson of the Audubon Society, this "balance" was a farce. As Peterson put it, "To put him in charge of the Interior Department is a crime."[28]

THE REACTION TO THE REACTION

The Reagan antienvironmental revolution represented the most staggering and comprehensive peacetime rollback of environmental policy in the history of the conservation movement. But it also helped to unite and energize a new coalition of liberal, pro-environment, anti-Reagan dissenters—a coalition that paired professional environmentalists with a wide variety of scientists, including many climate scientists, and a vocal minority of congressional Democrats. Already, *Global 2000* and the associated issues of acid rain, ozone depletion, and deforestation had sparked a hesitant cooperation between these groups in dealing with global environmental problems. It was their common response to the Reagan reaction, however, that brought these groups together to make global warming into a mainstream, Democratic, environmental issue in the 1980s.

Political dissent in the early 1980s took many forms; but especially for congressional Democrats, the first line of defense came in the familiar form of congressional hearings.[29] In the wake of Reagan's resounding electoral victory and his subsequent whirlwind of deregulation activity, Democrats hoped to curb the new administration's political momentum and set parameters for the president's popular mandate by publicly challenging Reagan's appointed officials and drawing out the implications of the administration's budget requests.

For the most part, it was politics as usual. As Senator Paul Tsongas of Massachusetts noted in Watt's confirmation hearings, few members of Congress realistically expected to block Reagan's appointments or significantly alter his administration's approach to environmental regulation. "Being confirmed is a piece of cake," Tsongas told Watt. "You will be confirmed.... That is not my doing."[30] Tsongas and his colleagues simply sought to extract as high a political price as possible for their eventual approval of Reagan's initiatives by undercutting the credibility of his administrators and publicly questioning the wisdom of his policies.[31] Technically, whomever Reagan appointed to administer federal agencies served not only the president but also Congress. In the early 1980s, Democrats used hearings to rein in Reagan's aggressive policymakers by reminding them—and the American public—of their dependence on Congress.[32]

The hearings also provided a forum for professional environmentalists to voice opposition to Reagan's appointments and environmental policies, a complement to mobilizing their constituents against the new president. Ironically, while Watt was a walking catastrophe for the environment itself, his objectionable and often reckless antienvironmental rhetoric helped spark a revival of America's environmental movement. In the midst of their political and policy defeats at the hands of Watt and the Reagan administration, environmental organizations of all stripes—alongside traditional liberal interest groups like the American Civil Liberties Union and the National Organization for Women—saw their memberships mushroom. In the aggregate, these organizations experienced a 33 percent increase in fund-raising returns in 1981, in large part from the combination of direct-mail campaigns and press coverage of Watt's confirmation hearings.[33]

The resurgence spoke to the ongoing realignment of the politically fragmented environmental movement with the mainstream Democratic left. The traditional conservation movement had deep Republican roots; but as David Brower of Friends of the Earth and former Republican governor of Delaware and head of the Audubon Society Russell Peterson intimated in the Watt hearings, the new administration's policies threatened to accelerate a recent trend away from the Republican Party among the conservation movement's rank and file.[34] Meanwhile, as leaders within more liberal organizations like the Sierra Club and Natural Resources Defense Council (NRDC) saw friendly old-guard Republican conservationists like Russell Train and William Ruckelshaus increasingly marginalized in favor of administrators like Watt and Gorsuch, they too moved steadily to the left.[35]

The confrontational Watt hearings set the tone for Democrats' congressional response to the Reagan antienvironmental revolution, but it was the administration's energy policy that became the shared focal point for Reagan's political, scientific, and environmental critics. Led in the House by Don Fuqua of Florida, James Scheuer of New York, Harold Volkmer of Missouri, George Brown of California, and Al Gore of Tennessee, Democrats focused on three main problems with Reagan's plan to restructure the Department of Energy and cut the agency's funding. First, they sought to demonstrate that dismantling the DOE would not save taxpayers money, as the administration claimed. By the Democrats' accounting, destroying the agency would actually cost the government money in the long run. Second, they went to great lengths to show how a breakup of the DOE would leave the government utterly unable to oversee, regulate, conduct research on, or even effectively monitor the production and consumption of energy in the United States. Finally, Gore, Brown, and Scheuer took the lead in emphasizing the potential environmental, economic, and national defense consequences of dismantling the DOE's research and development budget. It was in this context that Gore and his colleagues reintroduced the issue of CO_2-induced global warming.

A year earlier, in April of 1980 (when CO_2 had been just under 341 ppm), Brown, Scheuer, and Tsongas had held hearings before the House Committee on Energy and Natural Resources about CO_2 buildup in the atmosphere.[36] The hearings—which included NRDC cofounder and former CEQ chair James Gustave Speth, biologist George M. Woodwell of the Brookhaven National Laboratory, Wally Broecker of the Lamont Observatory, NCAR's William Kellogg (from the supersonic transport debate and SCEP and SMIC reports), and former CEQ member Gordon MacDonald, then of the Mitre Corporation—provided a sneak peek into the climatic conclusions of the forthcoming *Global 2000 Report*. Politically, Brown, Scheuer, and Tsongas hoped to use CO_2-induced warming as an example of the long-term environmental consequences of a Carter energy policy then tilting away from conservation and renewables and toward coal and synthetic fuels.[37] But for the most part, the hearings were soft on politics and proceeded amicably, the only tension a product of an ongoing scientific controversy between Woodwell and Broecker.

When Gore and Scheuer convened hearings on CO_2 and climate before the House Committee on Science and Technology in 1981, however, the tone of the discussion took what one scientist later described as an "ugly turn."[38] The hearings began benignly enough, with relatively uncontroversial testimony from Roger Revelle, Joseph Smagorinsky, Stephen Schneider, and Lester Lave, each of whom described the potential dangers of global warming and outlined areas of research that would generate the information Congress needed to make effective policy.[39] In part, the proceedings reflected Gore's and Scheuer's genuine, continued concern over an increasingly disconcerting environmental issue—an issue that Gore would make a centerpiece of his long political career.[40]

But the congressmen also mobilized these scientists to demonstrate the Reagan administration's myopia and even negligence in cutting government funding for CO_2 research at the DOE. Reagan's aides had kept the details of their DOE budget reallocation quiet, but Gore and Scheuer—along with the scientists they invited to testify—knew that administration officials planned to cut CO_2 research along with the majority of DOE's other environmental research programs. In the second half of the hearings, Gore and Scheuer called on Reagan's new appointments in the DOE and Office of Science and Technology Policy and challenged them to defend their yet-to-be-released budget cuts, not only in the face of state-of-the-art climate science, but quite literally in the face of the scientists who had produced it.[41] The confrontation made the administration look bad.

The Democrats' public approach to CO_2 and climate yielded three important results. First, Gore's 1981 hearings, along with subsequent hearings in 1982 and 1984, helped establish both personal and ideological relationships between prominent climate scientists (many of them already relatively liberal) and the mainstream political left.[42] Democrats had sponsored hearings on CO_2 and climate in the past; but unlike in the informational hearings of the 1970s, the confrontational nature of the hearings of the early 1980s all but forced scientists to take sides. Reagan's appointees—in particular, N. Douglas Pewitt, assistant to Energy Secretary James Edwards—alienated scientists like Revelle and Schneider by threatening to cut their DOE funding. In the process of defending the cuts, these appointees directly challenged both the value and substance of scientists' work.[43] Gore, Brown, and Scheuer provided an attractive alternative. Most obviously, the Democrats called hearings to give scientists a voice in defending funding for CO_2 research. They were natural allies. But Brown, Scheuer, and even more notably Gore, a one-time student of Revelle's at Harvard, also demonstrated a high level of scientific literacy and a genuine concern about global warming that impressed the scientists on a personal level. Gore's Academy Award–winning 2006 global warming exposé, *An Inconvenient Truth*, bears testament to the depth and longevity of this concern.

Second, in these hearings, Gore helped to publicly establish CO_2-induced climate change as an important issue associated with the ongoing debate over energy.[44] "If we take CO_2 seriously," Lester Lave of the Brookings Institution told the *New York Times* in 1981, "we would change drastically the energy policy we are pursuing."[45] The 1981 hearings failed to save the DOE's CO_2 program from the budgetary meat-ax, but the press picked up the relationship between CO_2 and energy and ran with it. "The threat of an overload of carbon dioxide, changing the planet's climate, is perhaps not immediate," read a *Washington Post* article. "But it would impose an absolute limit on fuel consumption, making the oil crisis and dislocations of the past seem trivial."[46] Since the late 1970s, Gordon MacDonald had argued that CO_2 should be incorporated into energy research. In the 1980s, Gore and his Democratic colleagues ensured that CO_2 had a place in energy politics.

Finally, in part because of the focus on energy, Gore's hearings all but demanded that American environmental organizations enter the public discussion about global warming. "There is of course no way that researching CO_2 build-up cannot call into question the current Administration's energy policy," wrote Anthony Scoville in the Friends of the Earth newsletter. "The Administration is cutting funding into CO_2 research so that it can justify its deliberate ignorance of the issues at hand."[47] Rafe Pomerance outlined the organization's new position on what he called "one of the most serious and irreversible environmental problems yet faced by man" in testimony submitted to Gore's 1982 CO_2 hearings: "Climate change must now be included as part of energy and economic innovation policy. Faced with the possible threat of global warming it is irresponsible to decimate programs for energy conservation and solar energy as the Reagan Administration is doing."[48]

The Environmental Defense Fund, though it did not participate in the hearings, soon decided that the issue was sufficiently important to hire Harvard atmospheric physicist Michael Oppenheimer as a senior scientist.[49] The NRDC continued to incorporate CO_2 into its international policy positions, and Gus Speth's new organization, the World Resources Institute, made climate change one of its foundational priorities. It would take until the second half of the decade for environmentalists to coordinate their efforts on CO_2 and climate, but Gore's hearings ensured that when global warming finally began to stick as a mainstream issue in the

summer of 1988, environmental organizations—and their constituents—
were ready to act.

INSIDE THE FEDERAL BUREAUCRACY

The fight over funding for CO_2 research extended beyond the halls of
Congress; government scientists also fought to maintain control over climate research within the federal bureaucracy itself. Again, discussions
about CO_2 provided a platform for criticizing Reagan's energy and environmental policies, but in this arena the results were more mixed. If the
Reagan administration did one thing well, it was controlling the federal
bureaucracy.

The first important conflict between the administration and its own
scientists occurred at NASA's Goddard Institute for Space Studies, headed
by James Hansen, and not surprisingly it again involved a dispute over
DOE funding. A taciturn Iowan and at that time the picture of a careful,
plodding scientist, Hansen had spent much of the late 1970s helping to
design a new general circulation model that would reveal the long-term,
average characteristics of climate independent of the short-term, ampli-
tudinal characteristics of weather.[50] Funded by an internal NASA grant
under the Carter administration, Hansen's group used the model—called
Model Zero until 1981, when GISS updated and renamed it Model II—to
explore the earth's climatic response to a doubling of atmospheric CO_2
given a variety of feedback processes and systemic sensitivities. Unlike
with other, higher-resolution GCMs, Hansen's group could run the new
model multiple times relatively quickly, allowing them to establish high-
and low-end boundaries for the potential impacts of certain poorly understood characteristics of the oceans on the overall climate prediction.[51]
Using historical climate data from as early as 1880, the model reliably produced results "roughly consistent" with measured reality.[52]

When Hansen finally published the Model II results in *Science* in 1981,
his conclusions did not go over well with an administration bent on downplaying the CO_2 issue. "The global warming projected for next century is
of almost unprecedented magnitude," Hansen and his colleagues wrote.[53]
The warming could lead, among other things, to the relatively rapid collapse of the West Antarctic Ice Sheet, producing an overall sea-level rise of
up to six meters—enough to "flood 25 percent of Florida and Louisiana, 10

percent of New Jersey, and many other lowlands throughout the world."[54] The paper also specifically referred to the need for a more diversified national energy portfolio in light of the potential climatic impacts of CO_2 from fossil fuels. "The degree of warming will depend strongly on the energy growth rate and choice of fuels for the next century," the GISS scientists wrote.[55] Their recommendations aligned with those of Carter's CEQ. They ran directly counter to the new president's energy policy: "CO_2 effects on climate may make full exploitation of coal resources undesirable. An appropriate strategy may be to encourage energy conservation and develop alternative energy sources, while using fossil fuels as necessary during the next few decades."[56] Worse still for the new administration, Hansen gave a copy of the paper to science journalist Walter Sullivan.[57] For *Science*, Model II represented the cutting edge of the continually improving world of GCMs. At the *New York Times*, a six-meter sea-level rise made front page news.[58]

The administration responded quietly and predictably: officials downplayed the newspaper report and went after Hansen's funding. Hansen's 1977 NASA grant terminated during the Carter-Reagan transition. In 1980, amid uncertainty over the future of the GISS climate program, David Slade, then still in his role as the DOE's chief CO_2 man, committed the DOE to supporting Hansen's work within the context of the DOE-AAAS climate program. After Slade's departure, however, Hansen had to appeal to Frederick Koomanoff to secure funding—the same Frederick Koomanoff called by Gore and Scheuer to testify on behalf of the Reagan administration's cuts in CO_2 research at DOE. Hansen's model had drawn specific technical criticism from some in the scientific community, raising questions about the certainty of Hansen's conclusions though not the quality of his approach. Koomanoff played up these criticisms and used them as grounds for denying Hansen's application, effectively curtailing the GISS project.[59] The subsequent funding cuts forced Hansen to lay off five researchers.[60]

More insidiously, as Hansen told Erik Conway in 2006, the new DOE officials also warned him not to share Model II with would-be collaborators at Pennsylvania State University, suggesting that otherwise these researchers, too, might lose their government support.[61] It was not the last time Hansen would cross a Republican administration, nor was it the most notorious attempt by an administration to shut him up. In 2008, still working in the NASA offices in Columbia University's Armstrong Hall above Tom's Restaurant (made famous by the television series *Seinfeld*), Hansen famously came to loggerheads with the George W. Bush administration for its attempts to censor the results of climate science conducted at NASA. In retrospect, the Bush administration could not have had a better template for dealing with Hansen than the one Reagan's people provided in 1981.

Silencing Hansen was the first in a series of ad hoc steps by the Reagan administration to use the mechanisms of government bureaucracy to quash the fruits of Carter-era research on CO_2 and climate change. The effort included more than a reduction in research funding or the marginalization of individual scientists like Hansen; it also involved capitalizing on professional and political divisions within the scientific community in order to manage the public scientific message on CO_2. The administration used these tactics perhaps most effectively in handling a 1983 report on CO_2 produced by a National Academy of Sciences committee chaired by the director of the Scripps Institution of Oceanography, William Nierenberg.

By 1980, the climate science community had come to agree on a few basic principles vis-à-vis CO_2 and climate, but scientists from different disciplines and institutions continued to disagree over the timing and magnitude of a potential CO_2-induced warming. In 1979, an NAS study chaired by Jule Charney provided the first self-conscious scientific consensus on CO_2 and warming.[62] The researchers concluded that a doubling of CO_2 would lead to a 1.5–4°C increase in the global mean temperature—a warming that might have significant social and environmental impacts.[63] As scientific debate over Hansen's 1981 paper showed, however, unanswered questions about the behavior of the oceans and the biosphere within the overall climate system made it difficult to predict how quickly and how severely the real-world climate might change in response to CO_2. Though both the Charney Report and Hansen's paper identified a moderate recent increase in temperature associated with CO_2 buildup, the models revealed only the faintest signal of a climatic response to the nearly 15 percent increase in CO_2 that had occurred since the mid-nineteenth century.[64] Hansen argued that the earth would begin to feel the impacts of global warming early in the twenty-first century, but other scientists contended that the ocean-induced lag in CO_2's influence on the climate system could last a century or more, and still others all but discounted the lag altogether.[65]

The 1983 NAS study headed by Nierenberg was a follow-up to the 1979 Charney Report, mandated by the National Energy Security Act of 1980 signed by Carter. Nierenberg used these unanswered scientific questions about timing to cast doubt on the basic conclusions of climate science as a whole.[66] The 1983 NAS committee noted that the potential impacts of atmospheric CO_2 represented a "cause for concern," but overall the report downplayed the urgency of the issue and recommended a "wait, do more research, and see" approach to the problem.[67] Thomas Schelling of Harvard and William Nordhaus of Yale, the committee's two economists (a new element in the NAS approach to climate change), argued that in a world of slow and gradual climatic change, humans would likely learn to adapt as quickly as the CO_2 problem developed.[68] Tied to energy, pollution, and the global environment, the 1983 report said, CO_2 presented an intractable problem.[69] But as Nierenberg told the *New York Times*, the committee thought that "we have 20 years to examine options before we make drastic plans. In that 20 years we can close critical gaps in our knowledge."[70] With good scientists on the case, CO_2 was no reason to upset the economy through major changes in energy policy.

The 1983 Nierenberg Report backed off considerably from the urgency of the 1979 Charney Report, and its conclusions clearly reflected the influence of conservatives within both the NAS and the Reagan administration. The Energy Security Act mandated that the Office of Science and Technology Policy administer the NAS study, which gave George A. Keyworth III, the president's science advisor, control of the Nierenberg committee's funding through 1983.[71] Nierenberg himself, though selected as chair of the committee (after some lobbying) before the 1980 election, was a political conservative who later served on Reagan's 1981 transition team and as an advisor to the president on atmospheric issues throughout his two terms in office.[72] To the extent that the administration directly pressured NAS researchers to toe the line—which, with Keyworth and Nierenberg in charge, was not really necessary—that pressure was relatively subtle, in the form of a tacit budgetary threat from none other than Frederick Koomanoff. Koomanoff held the purse strings for what little DOE funding remained available for CO_2 research in the early 1980s, and he quietly made it known that he hoped the NAS recommendation would line up with the administration's low-key position on the climate issue.[73] Ultimately, the fragmented, 496-page, multiauthor Nierenberg Report expressed a variety of viewpoints on CO_2, but Nierenberg controlled the overall presentation and penned the summary himself. By framing the issue as a scientific problem of little immediate concern—and by undercutting the consensus of the 1979 Charney Report—Nierenberg gave the administration the best result it could have hoped for. This was science-first climate change advocacy turned on its head.

If the Nierenberg Report helped to strip the urgency from the CO_2 discussion, a competing report put out by Stephen Seidel and Dale Keyes of the EPA, *Can We Delay a Greenhouse Warming?*, enabled the Reagan administration to further undercut the certainty of climate science in the public eye.[74] Produced in a rush, the EPA report took the opposite approach to that of the NAS study. Seidel and Keyes described CO_2 as an urgent problem that required a reevaluation of national energy policy. To the extent that the impacts of CO_2 remained uncertain, they provided grounds for precautionary action, not inaction.

Despite their conflicting conclusions, both reports actually confirmed the inevitability of greenhouse warming, but George Keyworth and White House counsel Ed Meese played up the disparities between Nierenberg's "sober" NAS report and the "unnecessarily alarmist" EPA study, imbuing press coverage of the climate issue with a sense of confusion rather than one of concern.[75] The press, not surprisingly, took more interest in the "debate" between the EPA and NAS scientists than in the broader implications of the science itself. Both studies were soon forgotten.

NUCLEAR WINTER

Reagan's energy policy figured prominently in reshaping the political landscape of global warming in the 1980s, but the new politics of global warming did not revolve entirely around fossil fuels. They also entwined with less obvious political issues, the most important of which was national defense. The link among CO_2, defense policy, and disarmament became enmeshed in the debate over "nuclear winter," bringing the Cold War back into the politics of global warming in a new way.[76]

Coined by NASA's Richard Turco in 1983, the term *nuclear winter* described a dramatic decrease in global mean temperatures that they believed would accompany a major nuclear exchange. Nuclear winter had very little to do with CO_2 and nothing to do with energy policy, but the issue brought together nearly every major player in atmospheric science in the 1980s, and the ensuing controversy helped to shape both the science and politics of global warming in the years to come.

The nuclear winter debate first arose out of an unlikely mix of Cold War politics, paleoclimatology, and ozone research. During the 1980 presidential campaign, Reagan had stumped for a more robust and aggressive national defense strategy. He attacked Carter's halting attempts to cooperate with the Soviet Union on arms control as an attitude of "defeatism," and he called the still unratified Strategic Arms Limitations Treaty (SALT II) "fatally flawed."[77] He and his secretary of defense, Caspar Weinberger, argued that Carter's policies had opened up a "window of vulnerability" wherein the United States lacked the technological systems to respond to a Soviet first strike.[78] The administration called for "rearmament," an expansion and updating of the U.S. military arsenal and the means to deploy it.[79] Weinberger argued that rearmament would allow the United States to regain the nuclear parity at the heart of the policy of deterrence, but he also launched a series of studies on strategies for fighting—and winning—a protracted nuclear conflict with the Soviet Union.[80]

For foreign policy doves and American liberals already frightened by Reagan's increasingly belligerent Cold War foreign policy and push for rearmament, Weinberger's comments about "prevailing" in a nuclear conflict were a divergence from the objectionable but still useful concept of mutually assured destruction that underpinned deterrence. Their response included congressional hearings, public exposés on nuclear destruction like Jonathan Schell's 1982 *Fate of the Earth,* and, most notably, the grassroots nuclear freeze movement that led to large antinuclear demonstrations and legislation—which passed the House—resolving to cease the production of nuclear weapons.[81]

In the United States and abroad, environmentalists stood in solidarity with this resurgent antinuclear movement. Since as early as the 1972 U.N. Conference on the Human Environment—with its roots in Swedish concerns about nuclear weapons and their dominance in international politics—environmental activists around the world had associated wars and nuclear armaments with environmental destruction and peace with conservation.[82] Reagan's defense policies reawakened long-standing concerns over the environmental impacts of military activities. In 1981, for example, the Sierra Club's Task Force on the Environmental Effects of

Military Projects drafted a resolution recognizing the "unprecedented destruction of the global environment" at stake in a nuclear exchange and expressing "grave concern over the lack of progress in completing nuclear arms reduction agreements."[83]

Scientists began to build on environmentalists' concerns. In 1982, the Swedish Academy of Sciences asked Paul Crutzen, then at the Max Planck Institute for Chemistry, to investigate the potential effects of a nuclear exchange on the ozone of the upper atmosphere for a special journal issue on the environmental consequences of nuclear war.[84] Ozone had become a popular and important environmental issue both in Europe and the United States in the early 1980s, and the scientific journal *Ambio*'s editors recognized that a global nuclear war would inject enormous amounts of ozone-depleting nitrogen oxides into the atmosphere. Strong supporters of disarmament, the Swedish academy hoped to use this stratospheric impact to reinforce the global stakes of a U.S.-Soviet nuclear conflict.[85]

Crutzen did *Ambio* one better. In 1980, Nobel laureate physicist Luis Alvarez at the University of California, Berkeley, and his geologist son, Walter, had published a controversial study that attributed the extinction of dinosaurs to climatic changes resulting from ash and dust sent into the stratosphere by a ten-kilometer-wide asteroid some sixty-five million years ago.[86] Crutzen and his *Ambio* coauthor John Birks of the University of Colorado realized that though scientists had studied the immediate thermal impacts of nuclear blasts, nobody had yet accounted for the climatic impacts of the soot, smoke, and other particulate matter that a nuclear war and the resulting fires could distribute through the atmosphere. Crutzen and Birks's "Atmosphere after a Nuclear War: Twilight at Noon" considered the possibility that a nuclear war could produce climatic effects similar to those caused by the Alvarezes' asteroid. Its title dramatically printed over a picture of a mushroom cloud, the article also suggested that the consequences of this climatic change might again be the extinction of the earth's dominant species—humans.[87]

In the spring of 1982, a group of planetary scientists at NASA's Ames Research Center in California made up of Brian Toon, Tom Ackerman, and Jim Pollack took up the Crutzen and Birks hypothesis and began to work it into their own atmospheric models. Supported by Carl Sagan of Cornell University and Richard Turco of NASA defense contractor Research Associates, the so-called TTAPS group (Turco, Toon, Ackerman, Pollack, and Sagan) was already applying their knowledge of planetary atmospheres and terrestrial volcanism to both the controversial asteroid thesis and the climatic impacts of a nuclear exchange.[88] Inspired by the Crutzen paper, they stepped up their research and began to use their one-dimensional radiative-convective model, along with publicly available information on nuclear yields and damage estimates, to simulate the climatic impacts of nuclear exchange scenarios.[89] Designed to handle multiple scenarios relatively quickly, the simple model did not include the ocean (an important moderating force in the climate system, as the authors recognized), nor did it account for seasons or horizontal circulation. It was more like a black-and-white snapshot than a full-color video of the earth's processes. Nevertheless, the TTAPS results strongly suggested that even a modest nuclear exchange could, as Crutzen suggested, lead to major atmospheric changes, including a massive depletion of ozone and the sudden onset of subfreezing temperatures during the summer.[90] Forbidden by NASA to use "war" in the title of any published material, Turco labeled these cold, dark postapocalyptic conditions "nuclear winter."[91]

Initially, the TTAPS group took great pains to distance their work from defense politics. They feared that too close a connection to the ongoing controversy over the MX missile and Reagan's rearmament push might damage their scientific credibility and jeopardize their government funding. Leaders at NASA gave specific instructions for what the civilian agency could and could not support with money and computer time vis-à-vis defense-related research, and they made it clear that TTAPS stood very close to the edge. As Turco later recalled, "There was tremendous pressure for us to not get involved."[92]

Carl Sagan, however, was nothing if not politically involved.[93] A talented science popularizer and one of the more famous scientists alive in the United States at the time because of his hit public television series, *Cosmos*, Sagan hoped to capitalize on the political implications of the TTAPS findings to demonstrate the folly of rearmament. In 1982, representatives of the Rockefeller Family Fund, the Henry P. Kendall Foundation, and the National Audubon Society approached Sagan and asked him to help organize a conference on the long-term consequences of nuclear war. Sagan recruited Stanford biologist Paul Ehrlich, NCAR founder Walter Orr Roberts, and Brookhaven National Laboratory's George Woodwell. This steering committee soon secured financial support and political backing from a wide variety of scientific and environmental groups, including the NRDC, the Environmental Defense Fund, and the Sierra Club. Many of these supporters were the same scientists and environmentalists who had testified in Gore's global warming hearings in 1981, and they would return to global warming in the future. In 1982, however, led by Sagan, the steering committee made the TTAPS nuclear winter paper, then still in progress, the centerpiece of its efforts.[94]

Sagan's group began by working within the scientific community to establish widespread agreement on the basic phenomenon of nuclear winter. In April of 1983, with the draft of the TTAPS paper all but complete, Sagan invited one hundred prominent scientists from a variety of fields to assess the TTAPS model and its conclusions in a private workshop in Cambridge, Massachusetts.[95]

Sagan's nuclear winter meeting had three purposes. First, Sagan legitimately hoped to flesh out any fatal flaws in the theory or the data that might undermine the nuclear winter hypothesis before he introduced the controversial idea to a broader public. Second, by asking other scientists to comment on the work and incorporating their suggestions, Sagan hoped to win the backing of prominent scientists in case critics tried to undermine his policy position by challenging his science. Finally, the Cambridge workshop served as a challenge to other groups working on the scientific problem of nuclear winter—in particular, Stephen Schneider, Starley Thompson, and Curt Covey at NCAR and a group led by Mike MacCracken and Cecil "Chuck" Leith at the Lawrence Livermore National Laboratory—to move forward with their own research, which Sagan and his colleagues believed would strengthen the TTAPS case.[96] Despite discrepancies between model results and uncertainty among workshop participants about the basic assumptions of both nuclear defense strategy and atmospheric processes, Sagan saw this private review and the research that came out of it as a vote of confidence for the TTAPS hypothesis—and for the subsequent public conference.

With atmospheric scientists presumably on board after Cambridge, Sagan and his colleagues orchestrated the "Conference on the Long-Term Worldwide Biological Consequences of Nuclear War," held in Washington, D.C., to garner as much public and political attention as possible.[97] The steering committee scheduled the two-day event to begin on Halloween. On October 30, 1983, Sagan published an exposé on nuclear winter in *Parade Magazine,* a popular Sunday newspaper supplement with more than twenty million readers. Chaired by George Woodwell and kicked off by Stanford University's eloquent president, Donald Kennedy, the conference itself was less a scientific meeting than an extended, staged press release.[98] A satellite link—relatively new technology in 1983—connected an audience of several hundred scientists, journalists, and politicians to members of the Soviet Academy of Sciences in Moscow.[99] Sagan, drawing on a career in popular science, delivered a concise, accessible, and alarming synthesis of the TTAPS work, which he reinforced with a video, *The World of Nuclear Winter.*[100] Ehrlich, also a well-known scientist, not only for *The Population Bomb* but also for his frequent appearances on Johnny Carson's *Tonight Show,* then summarized a parallel study on the biological impacts of a nuclear exchange and a subsequent nuclear winter that he and his biologist colleagues had conducted.[101] On November 1, Sagan and Ehrlich followed up with an appearance on Ted Koppel's *ABC News Nightline.*[102] As intended, their "scientific congress" made headlines around the world.[103]

In Washington, the conference organizers insisted that they had "rigorously avoided drawing any policy implications from their findings," but Sagan had a detailed and specific policy response in mind, which he published in the journal *Foreign Affairs* shortly after the TTAPS study appeared in *Science* in December of 1983.[104] The TTAPS one-dimensional model identified a "crude threshold, around 500 to 2,000 warheads" for a nuclear exchange that could trigger a nuclear winter.[105] Sagan used this threshold to demonstrate the shortcomings of existing postures on nuclear defense and disarmament that failed to account for the possibility of nuclear winter. He addressed eight different defense strategies, from new treaties on targeting and yields supported by moderates and liberals to Reagan's proposed ballistic missile defense system (the centerpiece of the administration's Strategic Defense Initiative, or, pejoratively, "Star Wars").

Given that an exchange of as little as 10 percent of the U.S.-Soviet nuclear arsenals could trigger a climatic doomsday scenario, Sagan found all of these prospects lacking. "None of the foregoing possible strategic and policy responses to the prospect of a nuclear war–triggered climatic catastrophe seem adequate even for the security of the nuclear powers," he wrote, "much less for the rest of the world."[106] Moreover, he contended, in a world linked by the potential of climatic disaster, "beyond the climatic threshold, an increase in the number of strategic weapons leads to a pronounced *decline* in national (and global) security."[107] With the long-term effects of even a limited nuclear war promising mutual suicide, Sagan argued that only a build-down of nuclear stockpiles could restore a logically sound and politically credible policy of deterrence.[108] "For me," he wrote, "it seems that the species is in grave danger at least until the world arsenals are reduced below the threshold for climatic catastrophe."[109]

Despite its rhetorical punch, Sagan's *Foreign Affairs* article had little impact on the politics of disarmament. His ongoing efforts to parlay the science of nuclear winter into defense policy did have a profound effect on the politics of global warming, however. Not surprisingly, both the style and substance of Sagan's arguments irked conservative scientists. Many objected to the unorthodox presentation of scientific materials in *Parade* and *Foreign Affairs*, and they balked at Sagan's close association with Ehrlich, an outspoken liberal and environmentalist. Long-time Republican science administrators like Robert Jastrow and Fred Singer, along with one-time NAS president Russell Seitz, feared that activists like Sagan and Stephen Schneider might push the field into the wrong kind of politics. When *Science* ran an editorial with the TTAPS paper praising Sagan's and Ehrlich's work on the issue, conservatives both within and outside the scientific community were incensed at what they saw as a left-wing agenda at the journal. Partly in response to this perceived bias, Jastrow and a small group of colleagues, primarily scientists associated with the defense industry, formed the George C. Marshall Institute, a conservative think tank initially aimed at supporting Reagan's Strategic Defense Initiative and nuclear energy that soon became a vocal opponent of efforts to curb ozone depletion, acid rain, and global warming.[110] Meanwhile, Singer and Seitz publicly attacked Sagan's policy recommendations, not only by challenging the TTAPS group's scientific results but also by impugning their methodology and openly questioning their motives.[111]

Sagan anticipated these objections, and for the most part he and his colleagues responded to them professionally and convincingly.[112] But he did not anticipate the criticism from scientists who generally shared his political beliefs. His highly publicized disagreement with Stephen Schneider in particular further fed conservatives' claims that Sagan had exaggerated the TTAPS results. It also intensified the ongoing polarization of climate science in the 1980s.

Schneider began studying the nuclear winter issue with Curt Covey and Starley Thompson at NCAR in the spring of 1983, with funding from the Defense Nuclear Agency. In loose collaboration with the TTAPS group and the Lawrence Livermore National Laboratory, the NCAR team sought to create a more realistic picture of the climatic consequences of a nuclear exchange by re-creating the TTAPS scenarios using a three-dimensional climate model. Scientifically, the episode recalled Schneider's own paper on aerosol cooling with Ishtiaque Rasool in the early 1970s. Whereas the TTAPS study derived its conclusions from a model of a vertical column of "dead air," this time it was Schneider and his group who worked to incorporate seasonal variations, horizontal mixing, and the stabilizing force of the oceans into their model.

Almost immediately, NCAR's results began to differ from the TTAPS study. Qualitatively, the NCAR model supported the nuclear winter phenomenon: in the event of a nuclear exchange, in all likelihood smoke, soot, and dust would lead to a catastrophic cooling of large portions of the earth. But the details of the three-dimensional model revealed a more complex and somewhat less stark range of specific climatic responses. A nuclear war in the summertime had nearly the same severity of consequences in both the one-dimensional and three-dimensional models, but a January strike in NCAR's model (a scenario impossible to simulate in the TTAPS model) yielded much tamer impacts. In all of the three-dimensional scenarios, the oceans significantly modulated the overall temperature change, especially near the coasts, and ultimately the climatic result that Schneider and his colleagues arrived at resembled the variability of a severe and unpredictable autumn more than a deep winter freeze. Perhaps most important, the variability revealed by Schneider's three-dimensional model—the "real world," as he called it—seemed to undermine Sagan's "crude threshold" for perpetual winter.[113]

Between the fall of 1983 and the summer of 1986, Sagan and Schneider engaged in a series of exchanges on the issue. Initially, Schneider privately tried to convince Sagan to back off of the threshold idea, which Schneider believed to be an artifact of the one-dimensional model. Schneider shared Sagan's general bent toward disarmament, and like Sagan he hoped that the climatic effects of a nuclear exchange would help demonstrate the insanity of the "winnable" nuclear war concept. An outspoken liberal who had as early as 1977 encouraged Sagan to use his popular appeal to take more of a political stand on scientific issues, Schneider certainly believed in using science as a political tool.[114] Moreover, he and Sagan from the beginning jointly opposed the Reagan administration's blasé response to the issue. In a dismissive report released in early 1985, the Department of Defense claimed to have already incorporated nuclear winter into existing defense strategy, and Weinberger went so far as to suggest that nuclear winter actually strengthened the case for the Strategic Defense Initiative.[115] Throughout 1984 and 1985, Schneider's NCAR group consequently spent a good deal of time and ink defending the nuclear winter hypothesis and reiterating its potential policy impact, despite their problems with some of Sagan's science.

But for Schneider, politically motivated science had to be impeccable, and popular scientists like Sagan had a responsibility to alter the specifics of their political objectives as the science behind those objectives changed. For the forcing function of knowledge to work, the knowledge had to be the best it could be. When Sagan refused to budge on the threshold concept, Schneider felt he had to air his results publicly for the sake of scientific credibility. Over the next two years, the TTAPS group and Schneider's NCAR team publicly traded views in scientific journals like *Science* and *Nature*, in the popular press, and in the pages of *Foreign Affairs*. Schneider and Thompson argued that "the global apocalyptic conclusions of the nuclear winter hypothesis can now be relegated to a vanishingly low level of probability."[116] Nuclear winter thus could not provide the sole impetus for a build-down policy, as Sagan had described. Still, as Schneider pointed out, even nuclear autumn could have a catastrophic effect on world agriculture, and he continued to urge the defense community to more meaningfully incorporate the environmental consequences of nuclear war into strategic planning. For Schneider, like Sagan, this included "significantly reduced levels of arsenals."[117]

As with the 1983 EPA and NAS reports on CO_2 and climate, the disagreement between Schneider and Sagan on nuclear winter allowed skeptics and political opponents to cast the whole issue as speculative. Schneider repeatedly insisted that he and Sagan had "few fundamental differences," but conservative climate scientists, skeptical defense analysts, and the press highlighted the NCAR and TTAPS disagreements to foreground the uncertainties of the hypothesis and to undermine the credibility of the science.[118] In what has by now become a familiar refrain, conservative commentators echoed conservative scientists in accusing the greater climate science community of taking too cavalier an approach to the evidence.[119] Conservatives criticized liberals like Sagan and Schneider for going public with their results before they had fully fleshed out the science. As Edward Teller, a nuclear winter skeptic and a key booster of the Strategic Defense Initiative, commented in *Nature*, "Highly speculative theories of worldwide destruction—even of the end of life on Earth—used as a call for a particular kind of political action serve neither the good reputation of science nor dispassionate political thought."[120]

Though the hypothesis itself had little to do with CO_2, the nuclear winter debate brought together the increasingly intertwined science and politics of climate change. From the beginning, liberal scientists, environmentalists, and politicians gravitated toward nuclear winter because of its political implications. Much as they had mobilized CO_2 research to undermine the Reagan administration's energy policy, these same groups—composed largely of the same individuals—studied and publicized the climatic impacts of a nuclear war as a form of public resistance to Reagan's defense policy and to the administration more generally. Again, this dissent played out in congressional hearings (again sponsored by Al Gore), at scientific conferences and in scientific journals, and within the federal bureaucracy. To a much greater extent than the CO_2 issue did in the early 1980s, the nuclear winter debate also appeared in public media like newspapers, television, and magazines. And finally, again, the experience drove liberal climate scientists, Democratic politicians, and leaders of environmental organizations to cooperate in response to a common opponent: the Reagan administration. The organizers of the Halloween conference worked in Washington, D.C., offices provided by Gus Speth and his new World Resources Institute. The NRDC, meanwhile, worked to obtain documents on the U.S. Navy and Defense Nuclear Agency responses to nuclear winter—documents the organization distributed to both liberal scientists like Schneider and to members of Congress.[121]

Studying the climatic impacts of nuclear war provided climate scientists with a more subtle way of resisting the administration's position on CO_2-induced climate change as well. The nuclear winter debate gave many of the same scientists whose DOE funding had been cut under Reagan new access to government money through the Department of Defense (DOD).[122] The DOD-sponsored research paid scientific dividends, especially for climate modelers who used nuclear winter as a test case—a "digital thought experiment," as Erik Conway calls it—for their rapidly developing models of atmospheric radiation and circulation.[123] The ongoing controversy fostered a new level of interinstitutional collaboration that contributed to the development of the Community Climate Model (so named for its nearly universal usage), and scientists' experiences working with the model generated new questions not just about global circulation, climate sensitivity, and climate forcing but also about the impacts of these phenomena—an area of research specifically targeted in the DOE cuts.[124] The new DOD money thus helped climate scientists circumvent Reagan's funding cuts and continue climate research—research that ultimately provided the raw material for activist scientists' attacks on the very administration that had cut the DOE research budget in the first place. Nuclear winter research was in this sense not just a form of dissent but also a subtle form of retaliation.

SCIENTISTS, ENVIRONMENTALISTS, DEMOCRATS

Alongside the ongoing debate over the effects of CO_2 on climate, the nuclear winter saga at once reflected and helped shape a new, more combative landscape of climate change politics. During the 1980s, scientists began to tell the story of global warming in new contexts that gave CO_2 new political meanings for new groups of people. Scientists themselves continued to focus primarily on producing more and better scientific knowledge and on advocating for science policy and funding that would make their research possible. But in the 1980s, their science advocacy led them to engage in political conversations about the Reagan administration's policies on energy, environmental protection, and nuclear defense that attracted environmentalists and congressional Democrats also opposed to administration policies. Global warming advocacy continued to revolve around scientific debates and to unfold in the language of science; but when scientists joined with environmentalists and Democrats on issues involving energy and defense, they raised the political stakes. As a result, the science itself became more polarized and more directly political.

Reagan's aggressive conservatism—the Reagan reaction—galvanized disparate groups of scientists, environmentalists, and politicians into a single community of global warming advocates, and in doing so it also brought together three key themes in the history of global warming in a new way. First, there was the challenge of making a problem as nebulous and geographically diffuse as global warming into a meaningful issue for professional environmentalists and their middle-class constituents. Global warming advocates did not solve this problem in the 1980s; they inverted it. Charles David Keeling articulated the complexity and scale of CO_2-induced climate change in 1963, but in the political context of the 1980s, the diversity of carbon dioxide's sources and the ubiquity of its impacts enabled global warming advocates to tie CO_2 to a number of issues that Democrats' and environmentalists' constituencies cared deeply about. As a stand-alone issue, global warming still did not carry much political weight, but it made a very appealing tool of political dissent when attached to these other issues.

The same versatility that enabled environmentalists and members of Congress to wield CO_2 as a political tool also helped to break down the barriers between global warming and mainstream environmentalism, a second major theme of this history. Beginning with the response to the Reagan reaction in the early 1980s, scientists concerned about climate change would begin to ally with professional American environmentalists in an increasingly concerted effort to combat global warming. To the extent that they forged their relationship through congressional hearings on energy and environmental policy and within the debate about nuclear winter, the marriage of these two groups also reflected the influence of a third major structural force in the history of global warming: the Cold War. The particular way in which these three themes coalesced in the 1980s helped to shape the most important overarching feature in the political history of global warming: the primacy of science in global warming discourse.

It is important to avoid painting the changes in climate science advocacy in the 1980s with too broad a brush. As Schneider's personal struggle with the tension between his political beliefs and his professional responsibility reveals, atmospheric scientists were deeply ambivalent about the relationship of science to policy. That ambivalence persists in memory. Every scientist's experience represented a different mix of personal and professional values, and few remember the controversy as a pleasant time in atmospheric science. Careers were damaged and friends were lost. The battle to incorporate the results of climatic research into energy and defense policies certainly divided the community, but the contours of those divisions were neither static nor clean.

Still, within the intense individual responses to the nuclear winter and CO_2 debates lies a common recognition that between 1979 and 1985 the politics of atmospheric science fundamentally changed. In the context of the Reagan reaction, the Keeling Curve began to take on new political meaning for new groups of people, including not only scientists but also politicians, administration officials, and professional environmentalists. The politics of global warming became more partisan and more combative. The tone of the climate debate took this "ugly" turn in the 1980s in part because both scientists and politicians began to recognize what was at stake. Scientists worried not only about the social and environmental consequences of global warming but also the potential political and economic consequences, especially as they applied to the problems of energy.

And yet, while scientists, environmentalists, and Democrats used the environmental impacts of CO_2 to criticize and embarrass the Reagan administration for what they characterized as irresponsible positions on science research and renewable energy, no group offered specific policy solutions to the problem of global warming itself, at least not beyond the conservation efforts they were already advocating. Despite the international environmental community's best efforts to create a framework for handling global environmental problems, no legal or regulatory mechanisms existed for dealing with the global causes or consequences of climate change. The political resistance in the United States to even the suggestion that the fossil fuel energy mix might need to change did not bode well for more detailed solutions in the future, either domestically or abroad. Especially after the *Global 2000 Report*, groups like the NRDC, the World Resources Institute, the Sierra Club, and Friends of the Earth began to recognize that alongside acid rain and ozone, global warming represented both an ecological and a political game changer. With an unfriendly administration at home, these groups looked toward scientists and the international political community to help define how this new game would be played.

6

THE IPCC AND THE PRIMACY OF SCIENCE

IN MARCH OF 1988, GLOBAL ATMOSPHERIC CO_2 OSCILLATED THROUGH 352 ppm on its way to its May peak, this time at 354 ppm. With little fanfare, the Keeling Curve had turned thirty. For three decades, the Mauna Loa Observatory had collected and compiled measurements of atmospheric CO_2 that, when graphed by monthly mean over time, showed a 10 percent increase in the gas over that time period. In 1988, those 357 monthly averages had clearly taken the shape of the now-familiar undulating, upward-sloping curve associated with global warming. Annual oscillations—iterations of a story of planetary respiration—taken together constituted a larger secular trend in CO_2, the greenhouse gas most associated with the increasingly alarming issue of global warming.

The primary structures of the history of global warming had also begun to take shape by 1988. The impacts of rising CO_2 had not yet begun to appear, but the story already had the markings of tragedy. The science-first advocacy that scientists and environmentalists developed to combat global warming had failed to achieve meaningful political or environmental objectives. In fact, the top-down, science-first approach had actually begun to undermine these objectives by leaving scientists vulnerable to political change.

In 1988, a new iteration of the CO_2 story institutionalized this science-first approach to advocacy at the international level in the form of the Intergovernmental Panel on Climate Change, or IPCC. The IPCC represented the apotheosis of the "forcing function of knowledge," a self-conscious attempt to create a scientific consensus on climate change upon which to base international policy. The IPCC itself was also the product of a related set of iterative stories. Scientists and their associates in the United Nations modeled the IPCC on the remarkably successful efforts to tackle two other problems of the global atmosphere in the 1980s: acid rain and ozone depletion. The relationship among acid rain, ozone, and CO_2 was deeply problematic, however, and understanding how the effort to combat these two other environmental problems reinforced the primacy of science in global warming discourse is central to the larger tragedy of global warming.

THE VIENNA-MONTREAL PROCESS

For environmentalists, acid rain presented a much more familiar set of problems than did either ozone depletion or global warming. Though newly pressing at a global scale, it was not a new issue. In an 1872 book called *Air and Rain: The Beginnings of a Chemical Climatology*, the English chemist Angus Smith described the influence of airborne matter from coal combustion and the decomposition of organic matter as "acid rain," noting its effect on the chemistry of rainwater throughout the English, Scottish, and German countrysides.[1] In the 1950s, limnologists, agricultural scientists, and atmospheric chemists took up the issue as part of independent efforts to expand basic research within each discipline. By the late 1960s, the Swede Svante Oden had begun to incorporate these studies in a developing hypothesis that linked the causes of acid rain—industrial emissions of sulfur dioxide (SO_2), nitrogen oxides (NO_x, often pronounced "nox"), and other chemicals—with the phenomenon's environmental and public health impacts.[2] Oden demonstrated that European SO_2 emissions adversely affected Scandinavian plant growth, freshwater fish populations, and overall water quality. His work was part of what prompted Sweden's leadership in the U.N. Conference on the Human Environment in 1972.[3]

Oden's studies of acid rain helped to illuminate the potential geographical disparity between the sources and impacts of pollutants like SO_2 and NO_x. As with global warming, the issue required new thinking about transboundary environmental governance, both in Europe and in North America. As early as 1970, the Nixon administration considered levying a tax on SO_2 tied to energy production from coal.[4] Agreements between European nations and negotiations between the United States and Canada yielded first bilateral treaties and eventually the more broadly international U.N. Convention on Long-Range Transboundary Air Pollution, proposed in 1979 and ratified in 1983. As Natural Resources Defense Council and World Resources Institute founder Gus Speth remembers, the long-distance impacts of these pollutants on agriculture, species and their habitats, and water quality challenged environmentalists to think about air pollution as more than just a local, urban problem.[5]

Legally, however, acid rain was little more than a variation on a theme, and American environmental groups initially sought to tackle the problem primarily by expanding existing legal mechanisms at home. Because of their impacts on public health and natural beauty (they reduced visibility), SO_2 and NO_x (smog) were forms of air pollution in their own right within the "antidegradation" framework of the Clean Air Acts of 1967, 1970, and 1977.[6] The identifiable impacts of these gases on water quality and plant growth via acid rain allowed environmentalists to capitalize on the provisions of the Clean Water Act of 1972 as well, increasing their options as they lobbied for lower emissions thresholds and prepared lawsuits against businesses, industry, and government agencies.[7] Meanwhile, these same identifiable impacts, once publicized, helped to mobilize grassroots support for litigation and lobbying efforts tackling everything from public health to the sanctity of America's wilderness areas. In 1990, the amended Clean Air Act finally incorporated an SO_2 emissions trading scheme—setting the stage for "cap and trade" approaches to CO_2—but even before then, acid rain gave environmentalists both legal and political traction on a large-scale atmospheric problem without necessitating a wholly new strategy of environmental activism.[8]

Acid rain linked environmentalists' ideas about regional air pollution with concerns over large-scale atmospheric change, but it was the degradation of stratospheric ozone that brought environmentalists around to the legal protection of the global atmosphere as a whole. Ozone (O_3) is an allotrope—that is, a natural structural modification—of oxygen (O_2), present throughout the various layers of the atmosphere. In the lower atmosphere (the troposphere), ozone is both a pollutant and an effective greenhouse gas. High levels of ozone cause respiratory problems in animals and can "burn" the leaves of sensitive plants. In the stratosphere, though, where most of the world's ozone is created and destroyed, the gas absorbs ultraviolet energy from the sun, protecting the biosphere below from damaging wavelengths of light.[9]

In the late 1960s and early 1970s, concerns over ozone depletion revolved around the potential for water vapor and oxides of nitrogen from high-elevation supersonic transport flights to accelerate natural stratospheric ozone destruction. These concerns did not die with the end of the American SST program, however. As Paul Crutzen, Harold Johnston, and others studied the effects of stratospheric NO_x on ozone in the mid-1970s, they realized that other substances that seemed stable in the troposphere might undergo chemical changes as they interacted with radiation and ozone in the stratosphere.[10]

In 1974, Sherwood Rowland and Mario Molina published a groundbreaking paper that investigated the role of halogenated hydrocarbons (also known as halocarbons) in the destruction of stratospheric ozone.[11] The paper, which earned the two chemists the 1995 Nobel Prize in chemistry, addressed two specific chlorofluoromethanes (CF_2CL_2 and $CFCL_2$) commonly used as refrigerants and as propellants in aerosol spray cans. Molina and Rowland also suggested that chlorofluorocarbons (CFCs)and other halocarbons, like those in the widely used DuPont refrigerant Freon, might also catalyze the destruction of stratospheric ozone.[12] Scientists working in conjunction with the U.N. Environment Programme and the World Meteorological Organization soon created projections of CFC and halocarbon usage, from which they then attempted to extrapolate future ozone depletion.[13] By the time the *Global 2000 Report to the President* laid these issues out for policymakers in 1980, it was clear that halocarbons and CFCs presented a potentially significant danger to stratospheric ozone and, by association, to both the human and nonhuman denizens of the earth.

As an environmental issue, ozone depletion bridged the gap between regional air pollution, which was subject to existing legal and political frameworks, and global CO_2-induced warming. Ozone depletion and global warming shared two important attributes that distinguished them politically from acid rain. First, though *Global 2000* expressed concern over the expanding geographical reach of SO_2 and NO_x, the two gases' environmental and public health impacts directly affected the geographically discrete places where either the gas settled and people breathed it in or the chemicals fell as precipitation. CO_2 and stratospheric ozone, by contrast, only affected local human and environmental systems through their global action. The ozone "hole"—discovered as early as 1976 but only confirmed in the middle of the 1980s—had a vague geographical range in the Southern Hemisphere, but for the most part ozone depletion led to increased ultraviolet radiation (specifically, medium-wavelength UV radiation) reaching the earth's surface worldwide. It was thus a truly global problem.[14] Similarly, though climate change might affect different regions to varying degrees, depending on factors like latitude and existing marginal precipitation, CO_2 itself represented a danger only through its ability to impact the earth's total energy budget.

Second, as *Global 2000* underscored, both ozone depletion and global warming existed as problems primarily in the future tense. Most new environmental issues began with the recognition of obvious impacts, usually driven by direct observation. As early as 1852, for example, Angus Smith noticed that the sulfuric acid in the air in Manchester, England, caused the colors of textiles to fade and metals to corrode.[15] True concern over acid rain began with studies of the measured acidity of Scandinavian rivers and streams, and further investigation led to more robust scientific theory.[16] At least until the discovery of the ozone hole, by contrast, concerns over both ozone and global warming turned this process on its head. Scientific theory presaged future impacts, with few measurable data points to drive the problem home in the present.

There was also a direct political relationship between efforts to protect the ozone layer and the burgeoning fight against anthropogenic global warming, and this was probably the most important connection between the two issues. The international effort to combat global warming was built on the same institutional framework as the effort to curb ozone depletion.

As Gus Speth argues, the development of the 1987 Montreal Protocol on Substances that Deplete the Ozone Layer became the standard model for international policymaking on global environmental issues in the late 1980s and early 1990s.[17] In an ideal world, Speth explains, global environmental governance develops in four stages. First, scientists—in this case Molina, Rowland, Crutzen, and others—identify a problem and conduct research in order to produce an agenda for nongovernmental advocacy based on reliable scientific evidence. Second, after more research and "fact-finding"—conducted in the case of ozone primarily by UNEP and the WMO—the international community negotiates broad policy goals and agrees on the political procedures of a framework convention designed to accomplish these goals. The framework convention typically yields a treaty (the 1985 Vienna Convention) enforced through negotiated protocols (most notably the Montreal Protocol) that describe the more specific actions that individual nations must take to accommodate the treaty as the problem changes over time. Third, national-level political campaigns secure the convention's ratification by national governments, which then collectively codify the treaty under the rules of the United Nations. And finally, once in force, the protocols are implemented, monitored, and strengthened depending on their success over time.[18] As Speth writes, "The Montreal protocol is the crowning achievement of global environmental governance."[19] Indeed, it has succeeded remarkably in stemming the tide of stratospheric ozone depletion since it entered into force.[20]

The Montreal Protocol has served less admirably, however, as a model for international action on climate change. Unlike the buildup of CO_2, ozone depletion had a direct, tangible impact on human health—increases in the risk of skin cancer—that struck a chord with an environmental community already deeply engaged with the fight against cancer and keen on reducing environmental carcinogens.[21] The National Academy of Sciences estimated that the continued use of CFCs at 1974 rates could lead to a 14 percent reduction in stratospheric ozone. And in a world of reduced ozone, the NAS wrote, "all unshielded cells are highly vulnerable to sunlight and may be killed by relatively short exposure to full sunlight."[22] As *Global 2000* pointed out, a mere 10 percent reduction in stratospheric ozone "appears likely to lead to a 20–30 percent increase" in skin cancer.[23] "Ozone induced changes in UV radiation," *Global 2000* noted, "would change one of the conditions that has almost certainly influenced the evolution of life on earth so far, and … can be expected to precipitate a disturbance in the existing balance of life virtually everywhere on the planet."[24] Whereas climate change remained somewhere between a legitimate environmental concern and a scientific curiosity, environmentalists could treat ozone depletion as an environmental emergency.

Also unlike threats to the stability of the earth's climate, artificial threats to stratospheric ozone came from discrete and easily identifiable sources, mostly in the form of nonessential or replaceable consumer products like aerosol cans, Styrofoam, and certain refrigerants. Atmospheric CO_2, by contrast, came from fossil fuels burned around the world for energy. These fuels stood at the heart of the twentieth-century world economy, and unlike CFCs they had few ready replacements—and none at a global scale. Aerosol cans and halocarbon refrigerants were easy to vilify, especially for an environmental movement that had cut its teeth on more than a decade of consumer advocacy campaigns. The few large U.S. chemical companies that produced these products gave groups like the Environmental Defense Fund and the NRDC a target for the domestic legal strategies that had served them so well on other environmental issues. As early as 1976, a "Ban the Can" public relations campaign, alongside a legal push from the NRDC, led the EPA to mandate phasing out the production of aerosol products containing fluorocarbons, effective in December of 1978.[25] By the early 1980s, moreover, many of the companies that manufactured CFCs in the developed world had already found ozone-friendly substitutes, which they could adopt relatively easily to meet the voluntary standards of the Montreal Protocol.

Finally, the Montreal Protocol itself changed the international political landscape in which global warming advocates hoped to develop a convention on climate change. In creating a consensus on ozone, UNEP executive secretary Mustafa Tolba and his colleagues took advantage of political buy-in from the industrialized world, the eager support of a capable group of concerned scientists, and, perhaps most important, the element of surprise. Much of the process for building an international scientific consensus on ozone depletion and translating that consensus into an international legal regime unfolded in an ad hoc manner. International agencies like UNEP and the WMO, along with environmental and scientific NGOs like the NRDC and the International Council of Scientific Unions, carved out niches within this new legal framework as they developed it. The few companies and governments still opposed to regulating CFCs and other ozone-depleting gases in the 1980s found themselves decidedly on the defensive, reacting to these new forms of environmental advocacy.

The framers of the treaty on ozone hoped to help history repeat itself concerning CO_2. But history does not repeat itself. History informs the contexts in which events unfold and thus prevents its own repetition. The relationship between ozone and climate change is a case in point. During negotiations over a potential global warming treaty, opponents were able to take the initiative in part because of their experiences with the ozone debate. The success in regulating CFCs put powerful corporate and governmental bodies on notice. The Vienna-Montreal process—UNEP's explicit model for a treaty on climate change—gave energy-industry lobbyists and energy-hungry governments a rough outline of what to expect when it came to global warming. Opponents of regulation of greenhouse gases consequently were able to carve out their own niches alongside environmental NGOs in the negotiation process, which tipped the strategic playing field on climate change in their favor. Having watched Tolba and his NGO supporters capitalize on the momentum generated through the consensus process on ozone, leaders from energy-dependent and energy-producing industries joined the Reagan and Bush administrations to undercut talks on climate change every step of the way, both scientifically and politically. Thus, while the Vienna Convention and the Montreal Protocol helped put atmospheric change on the international political agenda, the very existence of these agreements upped the political ante on the climate problem. "Politics caught up with ozone," noted one of Tolba's key advisors during the effort to create a climate change treaty in the late 1980s, but "climate was born in politics."[26]

MECHANISMS OF KNOWLEDGE

Nothing reflected the political limitations of the Vienna-Montreal process as a model for climate change more clearly than the relationship between science and policy that emerged during negotiations over the first international treaty on climate change, the U.N. Framework Convention on Climate Change (UNFCCC). As with ozone, developing an international legal regime for climate change revolved around the production and validation of scientific knowledge. Initially, scientists sought consensus on climate change through processes of scientific assessment similar to those that had proved effective for ozone and acid rain. As scientists, environmental NGOs, and American and U.N. environmental agencies soon recognized, however, the form of independent scientific consensus that fostered political action on ozone proved insufficient for international regulation of CO_2 and other greenhouse gases. The governments of the United States, the Soviet Union, and Saudi Arabia, among others, as well as corporations like Exxon and Shell, understood that international scientific consensus on climate change presaged international political action, and they insisted on maintaining some control not just over the political process but also over the science behind that process. Without backing from powerful nations and corporations, leaders at environmental NGOs and the United Nations found that scientific consensus yielded very little real political action. If Mustafa Tolba and his colleagues hoped to replicate their success with ozone by using science to support an international convention on climate change, they first needed to establish political buy-in on climate change science itself. The mechanism they developed for this purpose was the Intergovernmental Panel on Climate Change.

By the early 1980s, climate scientists already had several mechanisms for creating independent scientific consensus. In the United States, the National Academy of Sciences had produced major assessments of the relevant components of climate, including *Energy and Climate* (1977), the Charney Report (1979), and the Nierenberg Report (1983).[27] The National Research Council went out of its way to avoid unsolicited policy proscriptions, however; and especially after the equivocal Nierenberg Report, few scientists saw the NAS as a viable forum for forward-looking discussions about global warming. Abroad, meanwhile, the International Council of Scientific Unions continued its decades-long collaboration with the WMO, and in the early 1980s the ICSU and WMO hosted a series of conferences on climate change in Villach, Austria, designed to thoroughly assess the problem. Like the NAS assessments, however, these ICSU-sponsored assessments self-consciously avoided the political implications of climate change. As they had in the past, the recommendations revolved around international research efforts and generally shied away from international politics.

That is, until 1985. Unlike its predecessors, the consensus report of the last of these Villach conferences, held in October of that year, contained somewhat vague but nevertheless striking recommendations for climate change policy. If CO_2 and other greenhouse gases continued to accumulate in the atmosphere, the report warned, "in the first half of the next century a rise of global mean temperature would occur which is greater than any in man's history.... Many important economic and social decisions are being made on long term projects ... based on the assumption that past climatic data ... are a reliable guide to the future. This is no longer a good assumption since the increasing concentrations of GHGs [greenhouse gases] are expected to cause significant warming of the global climate in the next century."[28]

In response, the Villach authors contended, "scientists and policymakers should begin active collaboration to explore the effectiveness of alternative policies and adjustments."[29] Spearheaded by Swedish meteorologist and long-time climate science advocate Bert Bolin at Stockholm University, a group of scientific leaders at Villach recommended that the ICSU, the WMO, and the UNEP modify the assessment process itself. These scientists wanted work to begin on an international treaty that would mitigate what they characterized as an increasingly dire global environmental problem. In other words, the consensus-making process should begin to explicitly address not just science but policy.

In 1986, Bolin formed an informal group of international climate scientists called the Advisory Group on Greenhouse Gases (AGGG) to do just that. Made up of Bolin and two members from each of three organizations—UNEP, the WMO, and the ICSU—the AGGG undertook to address the potential impacts and policy responses to climate change and to "initiate if necessary, consideration of a global convention."[30] In 1987, with a grant from the Rockefeller Brothers Fund, as well as support from the Environmental Defense Fund via Michael Oppenheimer and from George Woodwell and his Woods Hole Research Center, the AGGG launched a two-part conference focused on developing national and international policies for responding to climate change. Held in Villach, Austria, under the auspices of Stockholm's Beijer Institute in early October, the first meeting laid out the global and regional impacts of greenhouse warming and addressed "technical, financial, and institutional options for limiting or adapting to climatic changes."[31] The second meeting, held a month later in Bellagio, Italy, used the initial assessment as a platform for proposing regional policies to mitigate and adapt to climate change and suggested international institutional arrangements that could help implement these policies.[32]

In retrospect, the shift seems like a relatively minor one; but for climate scientists at the time, Villach and Bellagio marked a bold step in how climate scientists addressed policy. The meetings also introduced a new rubric for evaluating rising CO_2 based on the potential temperature increase and related sea-level rise it might cause. Bolin and his colleagues asked scientists to recommend quantifiable objectives based on what he saw as a strong consensus on the basic science ... and they did. The group suggested that international policymakers set goals that would limit sea-level rise to between 20 and 50 millimeters per decade and global mean temperature rise to 0.1°C per decade. They stated flatly that limiting global warming "could only be accomplished with significant reductions in fossil fuel use," and they weighed the costs of these reductions against the potential costs of doing nothing.[33] Most forcefully, they emphasized the immediate need for international regulatory mechanisms to limit greenhouse gas emissions, particularly for "an agreement on a law of the atmosphere as a global commons" or "a convention along the lines for that developed for ozone."[34]

Bolin and the AGGG brought their activist momentum into the 1988 World Conference on the Changing Atmosphere, held in Toronto, Canada, a mixed scientific and political meeting sponsored by former Canadian Meteorological Service head Howard Ferguson that dealt with both climate change and ozone depletion.[35] As they had in Villach and Bellagio, the scientists involved in the Toronto conference again emphasized the need for specific targets derived from past records of environmental change, and again they called for an international legal framework in which to secure these environmental goals. A group of energy experts at the Toronto conference called for a 20 percent reduction of global CO_2 emissions from 1988 levels by 2005—a quantitative stance that the conference ultimately adopted.[36]

For Bolin and the panoply of scientific and environmental NGOs present in Toronto, the Villach conference in 1985 had offered an authoritative consensus on the science of climate that could support the bold targets then presented at Villach and Bellagio in 1987 and Toronto in 1988.[37] Mustafa Tolba and his U.N. colleagues agreed, and with good reason. Participants at the 1985 Villach conference had come from universities and scientific organizations in twenty-nine countries, in both the developed and developing worlds, and their conclusions had reflected the state of the art in climate science. The resulting report expressed plenty of uncertainties, but then so had reports on ozone depletion in the fact-finding endeavors leading up to the 1985 Vienna Convention. By 1987, Tolba had already begun planning an international convention for climate change—modeled after the ozone convention—and he used the 1985 Villach conference as his scientific support.[38] NGOs like the AGGG and U.N. agencies like UNEP had little chance of implementing aggressive policies on energy and emissions without government backing, but scientists and some U.N. leaders thought that the scientific message from Villach was sufficiently definitive to garner support even from relatively conservative administrations in the United States and United Kingdom. Indeed, immediately following the conference Tolba appealed to Secretary of State George Shultz, who represented the largest producer of greenhouse gases and the most important player in any potential climate talks, urging the United States to take "appropriate" policy actions in preparation for a new convention.[39]

Few national governments were ready to accept the 1985 Villach conference as the definitive word on climate, however, least of all the United States. Domestic government agencies continued to disagree over the nature and magnitude of the problem, and the Reagan administration had no reason to support bold action. Perhaps more important, neither the DOE nor the EPA—the two main agencies responsible for climate change policy—had a place at the table in Villach, and the State Department would not accept an international scientific consensus that did not involve U.S. government scientists who could speak for the interests of the administration.

Not wishing to lose momentum for an international agreement, the policy board of the National Climate Program (NCP)—an "outpost in enemy territory" within the Reagan administration, as one scientist called it—came up with an alternative.[40] The NCP recommended a new intergovernmental body to oversee yet another comprehensive assessment of climate science, this one led by official government representatives rather than by independent scientists.[41] UNEP and the WMO soon parlayed this proposal into a permanent mechanism for building consensus on climate change: the Intergovernmental Panel on Climate Change. These organizations charged the IPCC with the express mission of creating a scientific consensus that could serve as the foundation for the U.N. Framework Convention on Climate Change.

Officially commissioned in 1988 and chaired by Bert Bolin, the IPCC rehashed much of the material discussed at Villach in 1985, but under a slightly different and more transparent—but also overtly political—assessment structure. Peopled primarily by scientists from government agencies and by midlevel diplomats, the IPCC consisted of three main working groups: Working Group 1 dealt exclusively with the state of the physical science of climate change; Working Group 2 discussed the potential economic, social, and environmental impacts of the phenomenon; and

Working Group 3 provided possible strategies for responding to the threat of climate change. In addition, a smaller, less formal fourth working group tackled the specific concerns and problems of developing nations, and an administrative bureau oversaw the larger process.⁴² UNEP and the WMO asked for reports from each group by 1990, at which point they planned to convene as a larger body to compile the reports into a single, authoritative assessment and to hash out an executive summary of that assessment.

The key to the IPCC was its intergovernmental character. Unlike at Villach, Bolin and his UNEP/WMO colleagues deliberately sought to engage national representatives in the consensus-making machinery of the IPCC. The IPCC consensus was to be more than an agreement among scientists or NGOs; it was to be an agreement among governments. In part, the intergovernmental nature of the new assessment was a response to U.S. criticisms of prior climate change meetings; indeed, the Reagan administration insisted that the consensus process involve official government scientists.⁴³ But independent of the United States, Bolin and Tolba also recognized the importance of establishing a process that gave international political actors ownership over the issue of climate change. By participating in the IPCC, even skeptical governments—the United States, the Soviet Union, and many countries in the developing world—tacitly agreed that the problem merited some sort of international solution.

Bolin spelled the argument out to an impatient Stephen Schneider in the late 1980s. Schneider complained that assessment work was bogging down real work in climate science; a new assessment process would only further divert resources away from original research and add little to the myriad national reports of the 1980s. "How many of those [assessments] are convincing people in India or Indonesia or developing countries who don't trust the science that comes out of it?" Bolin asked. "Will it be possible to have climate policy without having a scientific group in which various countries in the world have some political ownership?"⁴⁴ Once an inclusive IPCC released its conclusions, Bolin recognized, participating nations theoretically had little recourse to object about the basic facts when it came time to negotiate a framework convention on climate change. Scientists and U.N. leaders thus traded some control over producing consensus on the science of climate change in return for an implicit political commitment to cooperate on international mechanisms for climate policy.

The Montreal Protocol was not the only force shaping the context in which scientists introduced the Intergovernmental Panel on Climate Change. The IPCC was already in the works as the Keeling Curve quietly turned thirty in March of 1988, but a series of political, scientific, and environmental events that summer heightened the urgency and political significance of the assessment process, especially for the United States. Largely thanks to anomalies of weather (easily conflated with changes in climate), global warming hit the mainstream in 1988, and the issue's increased popularity raised the public profile—and the stakes—of both CO_2 and the science associated with it.

Throughout the mid-1980s, Al Gore and his fellow congressional Democrats continued to push the Reagan administration to begin dealing with the problem of climate change. In 1986, Senator Joe Biden introduced an initiative mandating that the president commission an executive-level task force to devise a strategy for responding to global warming—a strategy the president was meant to deliver to Congress within one year.[45] The initiative became the Global Climate Protection Act of 1987, which the president signed.[46] The bill in the end required very little real commitment from the administration, but it demonstrated an expanding congressional interest in climate change that raised the domestic political profile of the State Department's negotiations with UNEP and the WMO leading up to the IPCC. It also led to more congressional hearings on the issue, which helped keep global warming in the news and thus on the public agenda.[47]

Nothing put climate in the public eye more than the environmental events of 1987 and 1988, however. In the summer of 1987, a heat wave hit the eastern United States, followed by the onset of a severe drought in the major farm states of the Midwest. May, June, and July of 1988 brought even more heat to the eastern United States and Canada in what one author called "one of the most intense combinations of drought and heat since the Dust Bowl of the 1930s."[48] Crops in the Great Plains began to fail. The Mississippi River hit a record low water level, compromising the river's navigability. High winds and severe drought turned small fires in Yellowstone National Park into the largest conflagration in the park's recorded history, affecting 36 percent of the land within its boundaries.[49] Abroad, a bad hurricane season racked the Caribbean; floods ruined crops, killed livestock, and displaced millions of people in Bangladesh; and droughts in China and the Soviet Union threatened food supplies around the world.[50] Before these events, in 1986, Andrew Maguire of the World Resources Institute had testified before a Senate subcommittee that "mankind's activities are changing the atmosphere in ways that could profoundly affect the habitability of the Earth," and his comments echoed a chorus of similar sentiments from scientists and politicians.[51] In the midst of one of the warmest and wildest weather years on record, the media began to ask whether those profound changes had arrived.[52]

With the help of Colorado senator Tim Wirth, James Hansen of NASA's Goddard Institute for Space Studies—the same James Hansen whose research funding had been cut by the Reagan administration seven years before—took advantage of the weird weather to add urgency to a growing public concern. Wirth scheduled a hearing on global warming for June 23, 1988, in the heart of the summer and at a characteristically slow time for Congress. He invited Hansen and Sukiro Manabe of the Geophysical Fluid Dynamics Laboratory to testify. The temperature in Washington, D.C. reached 101°F.[53]

Hansen told the Senate Energy and Natural Resources Committee—and the media—that he was 99 percent sure that the world was warming; that scientists could ascribe that warming to the greenhouse effect, and, implicitly, to fossil fuels, with a "high degree of confidence"; and, perhaps most controversially among scientists, that the impacts of global warming could already be detected.[54] Hansen made it clear that no single warm day or heat wave, or even series of heat waves, proved the larger, longer-term phenomenon of global warming; but alongside colleagues Manabe, George Woodwell, and Michael Oppenheimer, he suggested that the unpredictable weather and frequent severe storms of 1987 and 1988 foretold the conditions of the climatic regime to come.[55] As Hansen most famously told Phillip Shabecoff of the *New York Times*, "It's time to stop waffling so much and say that the evidence is pretty strong that the greenhouse effect is here and is affecting our climate now."[56]

The media response helped turn global warming—and the environment more broadly—into a major political issue in the 1988 presidential election. Over the next year, climate change appeared on the front page of nearly every major newspaper, and magazines like *Time, Newsweek,* and even *Sports Illustrated* ran cover stories on global environmental problems.[57]

Time, which had dubbed Ronald Reagan "Man of the Year" in 1980, modified the award in order to name the Endangered Earth "Planet of the Year" in 1989.[58] Distancing himself from the Reagan administration's reactionary environmental policies, Vice President George Bush declared himself "the environmental candidate" during the 1988 presidential primaries (a bold statement considering that Al Gore was also in the field at that time, albeit for the Democrats).[59] Bush specifically tagged global warming as a new and important environmental challenge and vowed to counter the greenhouse effect with "the White House effect."[60]

Despite the increasing visibility of the problem, however, once in office Bush backed away from his commitment to fight global warming. His administration was slow to appoint a science advisor, and Bush received his main briefings on the issue from William Nierenberg, then working for the conservative Marshall Institute, which had recently published a report casting doubt on greenhouse warming.[61] Following Nierenberg's lead, the administration—and particularly White House Chief of Staff John Sununu—began to play up the uncertainties of atmospheric models and the climate system itself.[62] The president took what one historian has called an "America-first, business-first" approach to formulating policy on global warming.[63] The administration stressed the "likelihood and severity of economic risks that *would* be posed by regulation, while downplaying those that *might* be posed by 'potential' climate change."[64] Thus, despite unprecedented public interest and domestic political backing for addressing climate change, the Bush administration, like most other national governments, entered international negotiations on climate change in the late 1980s focused primarily on protecting national economic interests. Just as at home, within the international community the Bush administration used doubts about the science of global warming as a first line of defense against aggressive climate change policy. The IPCC became a vehicle for these doubts.

THE IPCC AND THE NEW POLITICS OF CONSENSUS

As the Bush administration understood, the overt ties between the IPCC and the forthcoming UNFCCC not only helped establish political buy-in on future climate change policy; these ties also raised the political and economic stakes of the IPCC's assessment of climate science itself. Whereas the 1985 Villach conference was just another in a series of warnings from the scientific community, Bolin and his WMO and UNEP colleagues meant for the IPCC to serve as a politically negotiated scientific baseline for binding international laws and policies. Both scientists and politicians already recognized that any serious effort to mitigate CO_2 and other greenhouse gases would involve major changes in national-level energy strategies and land-use policies—two sectors fundamental to economic growth in developing and industrial nations alike. But they also realized that the ability of any international legal framework to address climate change would depend largely on the level of certainty and concern expressed in the IPCC assessment. Consequently, debate over the language and details of the IPCC's first assessment report almost immediately became the front line of a larger battle over a future international legal regime—the UNFCCC—that many believed could have a significant impact on national economies and on the world economy as a whole.

For the U.S. and foreign governments alike, climate science was a surrogate for climate politics at nearly every level of discussion during the first IPCC meetings, beginning with the leadership and structure of the body itself. UNEP and the WMO allocated leadership of the working groups largely according to the political importance of the nations represented.[65] John Houghton of the United Kingdom headed Working Group 1 on basic science; the Soviet Union's Yuri Izrael took over Working Group 2 on impacts; and a U.S. State Department official, Fred Bernthal, served as the chief of the politically sensitive Working Group 3 on potential responses. Each chairperson worked with co- and vice-chairpersons from other nations so that each working group represented a constellation of different and often competing economic and political interests. Bernthal, for example, worked on policy responses with representatives from China, Canada, Malta, the Netherlands, and Zimbabwe; Houghton worked with representatives from Senegal and Brazil.[66] From 1988 to 1990, every step of the assessment process, from decisions about what questions to address and what scientists to consult, to discussions about what verbiage to use in presenting findings, reflected ongoing negotiations between the working-group chairs, cochairs, and vice-chairs. The IPCC quickly began to reflect both the beauty and terror that so often characterize international political negotiations.

Governments were not the only interested parties in the IPCC process. Once in session, working-group meetings also included scientific "observers" representing NGOs and industry. For environmentalists, the organization and influence of their business and industry counterparts was an eye-opener. In a May 1990 meeting of Houghton's Working Group 1 in Berkshire, England, for example, Jeremy Leggett, a Royal School of Mines geologist turned Greenpeace activist, remembers sitting in the back row alongside Dan Lashoff of the NRDC and "eleven scientists from the oil, coal and chemical industries, including two from Exxon, one from Shell and one from BP [British Petroleum]."[67]

These observers influenced the IPCC's assessment in two main ways. First, because the IPCC defined these observers' roles only loosely, they participated relatively openly in scientific discussions alongside government scientists, especially when those discussions dealt with the wording of the working group report summaries.[68] In Working Group 1, scientists like Leggett and Lashoff, who represented environmental organizations, consistently pushed the IPCC to explicitly spell out the worst-case scenarios of climate change for policymakers, to emphasize the full risk. Scientists representing the oil, coal, and gas industries, many of them employed by politically conservative think tanks committed to free-market ideals, downplayed these worst-case studies and emphasized the uncertainties of atmospheric modeling and the myriad, poorly understood complexities of the global climate system.[69]

Just as important, nongovernmental scientists from both industry and environmental groups sought to establish relationships with representatives of foreign governments in order to influence these governments' positions on controversial aspects of the IPCC reports. Here they acted essentially as lobbyists. Again, the two groups focused on attacking or upholding the validity of models and the certainty of the science more broadly.[70] And again, environmental NGOs found themselves outnumbered and overmatched by representatives from the energy industry, who were particularly effective in the role.

Negotiations over the specifics of the IPCC consensus only intensified as the working groups presented their reports to the IPCC plenary. In the working groups, scientists representing nations and political interests fought to define the boundaries of the information the IPCC would provide to policymakers. In the plenary, government representatives—including some heads of state—focused on protecting their nations' interests by adjusting the specific language in which the international community would present that body of knowledge to the world. European states fought to maintain the original language of the working-group reports in the final summaries, but the United States, the Soviet Union, and several other countries took the opportunity to revisit those reports with their own interests in mind. They sought to water down statements from Working Group 2 on the impacts of climate change and to highlight the uncertainties of the basic science laid out by Working Group 1. Few nations targeted Working Group 3 in the plenary, but only because that U.S.-led group had produced anemic and toothless policy proposals focused primarily on technological solutions, to which few nations could object.

The process was detailed and litigious, often maddeningly so. The United States and Saudi Arabia objected to the word "confidence" in the IPCC's proposed prediction on future warming.[71] The two nations also proposed replacing a call for understanding the costs of climate change with a call for understanding its "costs and benefits," and Saudi Arabia invidiously but understandably suggested replacing "carbon dioxide"— an inevitable product of their biggest export, oil—with the more general (and, to be fair, more accurate) "greenhouse gases."[72] Member nations of the Organization of Petroleum Exporting Countries (OPEC) argued that the final documents should explicitly clarify that the international community supported only "safe" nuclear energy solutions, thereby undermining the viability of a major energy competitor for oil.[73] India and other developing countries pressed for language pinning responsibility for CO_2-induced warming on the industrialized world, thus preemptively diminishing their own responsibility for future solutions under the UNFCCC.[74] Perhaps most famously, at a meeting in Geneva in November of 1990, Peter Timeon of the Micronesian Republic of Kiribati urged the group to say that the greenhouse effect and sea-level rise not only "could" but in fact "would" threaten the survival of island nations—a one-letter change that American representative John Knauss rejected, nearly scuttling the final plenary session of the IPCC in the final hour.[75] In each case, the debate focused on the scientific accuracy of proposed changes, but clearly the science stood proxy for the real political and economic interests at stake.

INSTITUTIONALIZING THE PRIMACY OF SCIENCE

The awkward marriage of science and politics under the IPCC underscored the dilemma at the heart of the fight against global warming. Scientists and environmentalists continued to believe that in the right political forum, good science would inevitably lead to political change, as rational political actors responded to the best information available. Successful efforts to translate scientific consensus on acid rain and ozone depletion into international policy action reinforced this optimistic notion. The high-stakes energy and land-use issues associated with potential solutions to climate change, however, exposed the political limitations of NGO and U.N. efforts to use scientific consensus to promote international regulations. Scientists realized that they needed strong political backing in order to implement political change on global warming. To convince the state officials and government agencies that could give their recommendations real political clout, environmental advocates sought to involve these political actors in the knowledge-making process. And to do this, they created a new mechanism for building scientific consensus on climate change: the IPCC.

Scientists and environmentalists alike lauded the completion of the IPCC's first assessment report as a transformative moment in international negotiations on climate change. It was an important feat. Just as UNEP and the WMO had intended, scientists and politicians had created a rigorous, transparent, and credible process for producing international scientific consensus on climate change, bringing the issue to international attention.[76] The international community recognized the IPCC as the first step in developing the U.N. Framework Convention on Climate Change, to be introduced at the 1992 U.N. Conference on Environment and Development in Rio de Janeiro. And late in 1990, the U.N. General Assembly accepted the first assessment report and officially commissioned the Intergovernmental Negotiating Committee to begin the process of translating the IPCC's scientific consensus into international policy. The working-group meetings, though technically not loci of new research, helped to push new developments in atmospheric modeling. The conversations of Working Group 1 in particular helped to expand discussions of greenhouse gases beyond CO_2 to include noncarbon gases like methane and ozone, in addition to other trace substances, that could affect the global climate system. On the whole, it is difficult to judge the IPCC fairly without recognizing its remarkable success.

At the same time, incorporating national interests and government and industry representatives into the international scientific consensus-making process had important and not altogether positive consequences for both the science and politics of climate change. The politically negotiated nature of the IPCC's consensus cut two ways. On one hand, participation in building consensus gave countries the political ownership over the issue that Bert Bolin and his colleagues had identified as essential to any long-term climate convention process. But on the other hand, for many scientists, national-level politicians, and even some IPCC participants, the overtly political nature the IPCC undermined the perceived "purity" of its consensus assessment. As science and environment scholar Stephen Bocking explains, "boundary organizations" like the IPCC are meant to "internalize within themselves the ambiguous border between science and politics, moderating the tendency of conflicting interests to dismantle the scientific claims of their opponents ... in effect stabilizing this border so that scientific claims are better able to meet the criteria of both political and scientific credibility."[77]

But in practice, scientists and politicians on all sides of the issue understood the negotiated nature of the IPCC process, and the consensus it produced carried only limited credibility within either the scientific or the political community. (The IPCC would later integrate peer review into its structure to protect the institution's scientific credibility, to limited effect). While opponents of action on climate change had recourse to lobbying and direct participation in the international scientific consensus-making process through the IPCC, once the IPCC released its assessment, skeptics from national governments and from the energy industry also had the option to simply deny the credibility of the institution itself.[78]

In addition, though many in the media and the public perceived the IPCC as a liberal organization making bold predictions on climate change for the sake of promoting world government, the IPCC's consensus-making process actually privileged a relatively conservative scientific position.[79] Scientific consensus requires scientists from a given field or fields to agree on a credible range of ideas or results that represent the state of knowledge on a given scientific problem. In an ideal world, groups of scientists would discard high- and low-end outliers equally. In reality, however, most scientists face a greater professional risk in overstating a case than in understating it, especially when the results imply political action. The size of the IPCC, the high stakes of the economic and political implications of its reports, and the multitiered structure of its negotiations in the working groups and the plenary only exacerbated this trend toward conservative results.[80] So, too, did the well-organized efforts of industry lobbyists and conservative governments serve to water down the IPCC conclusions. As a result, the first assessment report, though certainly an implicit call for policy action by its very existence, fell short of the bold recommendations that international scientific leaders like Bert Bolin had hoped for.

Finally, and perhaps most important, the IPCC allowed political actors to define the problem of climate change in terms of competing scientific views rather than the competing political and economic interests really at stake. UNEP and the WMO did not create the IPCC as a policy mechanism; they created it as a mechanism for producing a certain form of politically negotiated consensus knowledge. But because the IPCC overtly presaged the creation of the UNFCCC, delegates were able use this knowledge-making process to influence future political negotiations.

Negotiations by the Soviet Union—most likely one of the few "winners" in moderate scenarios of climate change—provide an apt example. As negotiations unfolded, the Soviets grew increasingly uncomfortable with the limitations a strong convention on climate change would likely impose on the development of its oil fields. Looking for wiggle room in future talks, Soviet representatives attacked the Working Group 1 consensus for failing to include Soviet scientists' paleoclimate data, which, though it did not undermine the broader IPCC consensus, created new wrinkles of uncertainty. The issue was clearly an economic one for the Soviet government, but the Soviets' primary tactic for defending their economic interests from potentially threatening climate change policy was to destabilize the consensus on climate change science.

The Soviets were not alone in their machinations. In Working Group 3, American officials attempted to shift the potential policy response to include not just CO_2 but also a wide variety of greenhouse gases the United States had already begun to regulate in other contexts.[81] Delegates argued over carbon sources and sinks, greenhouse gas concentrations, and the duration of those gases in the atmosphere, but ultimately these scientific debates were important only as far as they helped determine nations' liability under a future convention on climate change. The IPCC thus codified the consensus-making process as a de facto forum for geopolitics as usual, couched in the technical language of uncertainty and climatic complexity.

The technical language persists in twenty-first-century international climate negotiations. Between the first IPCC report and the unveiling of the UNFCCC, however, geopolitics as usual would change dramatically, and with it both international environmental politics and the politics of global warming changed as well.

THE GOSPEL OF THE MARKET

AT THE TURN OF THE TWENTY-FIRST CENTURY, JEREMY LEGGETT, THE geology professor and Greenpeace member who had found himself so overmatched by industry lobbyists during the first IPCC process, wrote an exposé on his role in international climate change politics. He begins *The Carbon War* on November 9, 1989, with an image seemingly unrelated to the global environment: the fall of the Berlin Wall.[1] For Leggett, the chaotic scene of positive change in Berlin represented more than the beginning of the end for Soviet-style communism. It also symbolized a broader positive change that, more than ten years on, gave him hope as he wrote his book. Nobody predicted the startling events of 1989 through 1991—not diplomatic historians, not political scientists, not even so-called Sovietologists in the U.S. State Department. If a political paradigm so dominant and long-standing could collapse so quickly and unexpectedly, then so too might the age of fossil fuel energy be rapidly and even peacefully brought to a close.[2]

In retrospect, Leggett's optimism seems misguided. If his "carbon war" has been won, Leggett and Greenpeace have certainly not been the victors. Even today, the concentration of CO_2 in the atmosphere, tracked in the Keeling Curve, continues to rise.

Still, the end of the Cold War remains a strikingly insightful place to begin a discussion about modern climate change politics. Climate science was born of the Cold War, and the politics of global warming emerged against a national and international political backdrop dominated by the conflict. This is one of the major themes of global warming's history. In the two years between the IPCC first assessment report in 1990 and the introduction of the UNFCCC in 1992, however, this dominant international political paradigm unexpectedly and quite suddenly vanished.[3] And just as the Cold War had shaped the key institutions and problems of climate science in the 1950s, so too did that conflict's sudden, unexpected, and mostly nonviolent end profoundly influence the politics of climate change in the early 1990s. The end of the Cold War changed the nature of the international system within which debate about CO_2 and global warming unfolded.

The different U.S. responses to two developments in 1990s international climate change policy underscore the impact of this changing foreign policy situation. The first development was the U.N. Framework Convention on Climate Change in 1992. A treaty signed by a moderate Republican president, George H. W. Bush, at the U.N. Conference on Environment and Development in Rio de Janeiro and quickly ratified by a Democratic Senate, the UNFCCC reflected the optimistic spirit of international cooperation that Leggett sensed in the immediate post–Cold War world. But the nonbinding nature of the treaty and the compromises leading to its ratification also laid bare emerging tensions in both the domestic and international politics of climate change. These tensions involved the priorities of domestic economic stability and environmental protection over both international development and the proper role of the United States in international environmental leadership.

Five years later, a second confrontation over binding targets in international climate change policy had a very different outcome. That confrontation occurred at a Conference of the Parties to the UNFCCC held in Kyoto, Japan, in 1997, and it revolved around exemptions from responsibility for carbon mitigation for the developing world. A Democratic U.S. vice president, Al Gore, salvaged the conference from deep divisions between Europeans and Americans over the role of markets in mitigating carbon. His colleagues in the Senate, however, resolved unanimously to reject any treaty that failed to include the developing world in a regime of CO_2 mitigation. The Kyoto Protocol was never introduced to Congress for ratification.[4] The rejection of the Kyoto Protocol stemmed in large part from changes in how politicians understood CO_2 and global environmental governance after the demise of the Cold War.

The end of the Cold War was more important, in more unexpected ways, to the politics of global warming than scholars have recognized. In part, the changes wrought by the end of the conflict represented extensions of some of the very forces that brought the Cold War itself to a close. One major change in climate change discourse—and, in fact, in environmental discourse more broadly—that coincided with the end of the Cold War was the rise of economics as the key rubric for environmental decision making. At its core was a new buzzword associated with both economic globalism and international environmental protection that would help define the politics of both the UNFCCC and the Kyoto Protocol: *sustainable development*.

Introduced in part as a foil for the pessimistic ethos of austerity characteristic of the 1970s, sustainable development was a notoriously slippery concept. It also has a contested history. The 1987 report *Our Common Future*, by the World Commission on Environment and Development (WCED) chaired by Gro Harlem Brundtland, brought the term into mainstream usage. "Sustainable development," the report explained, "is development that meets the needs of the present without compromising the ability of future generations to meet their own needs."[5] But for economists, the concept was nothing new. As economic historian Emma Rothschild notes, the concept of sustainability introduced in the so-called Brundtland Report was a variation on a much older concept of "maintaining capital intact," an economic principle fleshed out by A. C. Pigou, Friedrich Hayek, and John Hicks in the 1940s that cautioned against reducing capital stock for the sake of profit. Robert Solow and Ronald Coase each later expanded definitions of capital stock to include elements of the environment.[6] Even these early forays into environmental economics were not entirely new, however. As historian Paul Warde points out, German and Italian foresters and farmers have argued for the maintenance of natural capital in the form of soil conservation and silviculture since the 1600s.[7] The roots of sustainability as a concept, broadly construed, thus run much deeper than the Brundtland Report.

But "sustainability" generally and the "sustainable development" championed by the WCED were not the same things. The Brundtland Report was not just about the maintenance of capital. The WCED presented a particular version of sustainability geared toward reconciling environmental protection, social equity, and economic growth such that these three things became inseparable. That marriage—of equity, affluence, and environmental health—spoke as much to geopolitical goals is it did to purely economic ones. It was very much a product of the Cold War.

In *The Global Cold War*, Odd Arne Westad describes the ideological battle between Soviet-style communism and American capitalism as a conflict over the rightful succession to the mantle of European modernity—that is, over which system could claim the legacy of a culture and society defined by reason, progress, and justice. Westad argues that this conflict played out most visibly and importantly in the Third World—a term coined in the 1950s to refer to "the people" or the "global third estate," in Westad's words "the global majority who had been downtrodden and enslaved through colonialism, but who were now on their way to the top of the ladder of affluence."[8] Since the turn of the twentieth century, Americans had sought to export a package of economic aid, technological know-how, laissez-faire economics, and various forms of liberal democracy to Latin America, Africa, and the Middle East. These diplomatic and material overtures were meant not only to provide resources to impoverished states, but also to help establish new markets for U.S. goods and to demonstrate the desirability of the American way of life to nations making the transition out of colonial rule.[9] By the late 1950s, Soviet-style communism had supplanted other forms of totalitarianism as the ultimate foil for American capitalist democracy, and thus the battle for the hearts, minds, and markets of the developing world became a battle against the Soviets.

The key terms in the rhetorical battle for influence in the Third World were equity, liberty, and sovereignty. Communists prioritized equity—often framed as a fundamental issue of justice—but struggled to demonstrate their ability to provide the material wealth and consumer goods available to Americans. Americans, especially after the Second World War, rode the tide of liberty—buttressed by material wealth—but often the affluence they associated with liberty highlighted the gross inequalities that fueled Soviet criticism of laissez-faire capitalism.

Neither system did particularly well with the concept of sovereignty. Even as both systems promised to bring former colonies out of poverty and into a global middle class, both nations—the Soviet Union and the United States—relied on Third World natural resources and markets to sustain their own economies and to provide tactical and political advantages in the ongoing East-West conflict. Soviets and Americans each claimed that their respective political systems guaranteed Third World nations' sovereignty in different ways, but the quasi-colonial material relationship undercut such claims. Beginning in 1955, a number of developing nations, following the lead of powerful governments in India and Egypt, began to make a coordinated and concerted effort to at once capitalize materially on the East-West competition for Third World allegiance and reassert their political independence from the two superpowers. This "nonaligned movement" worked through the U.N. General Assembly to secure aid for economic development during the U.N. Development Decade of the 1960s. In the wake of this success, these same developing nations demanded that efforts to address global environmental problems at the 1972 U.N. Conference on the Human Environment in Stockholm also include concerns over economic development.[10]

Sustainable development was a Western, capitalist answer to those demands. As described in the Brundtland Report, sustainable development offered a way for the traditional organs of development aid—multinational banks, the International Monetary Fund (IMF), the World Bank, and other independent and nationally funded aid agencies—to at once distribute capital to promote the democratic capitalist system and to satisfy complaints about the equity of that system levied by left-leaning critics. Sustainable development also addressed concerns over environmental degradation raised both by Western nations and, importantly, by Third World citizens themselves. Sustainable development, the WCED claimed, should guide growth in all countries, "developed or developing, market oriented [code for capitalist] or centrally planned [code for communist]."[11] But the Brundtland Report made it clear that the concept would best thrive in a flexible, capitalist world. As the Soviet Union's ability to influence Third World development declined in the early 1980s, the sustainable development paradigm came to serve both as a mechanism for institutionalizing the triumph of the global capitalist system and as a check on that system's excesses. It also became the dominant framework in which to understand potential solutions to the problem of global warming.

RIO AND THE END OF THE COLD WAR

The concept of sustainable development entwined with the politics of global warming in the run-up to the most important event in international environmental politics since the 1972 Stockholm Conference: the 1992 U.N. Conference on Environment and Development (UNCED), which became known as the Rio Earth Summit. The Earth Summit was built upon the concept of sustainable development articulated in the Brundtland Report, and one of the event's major objectives was the introduction of an international treaty on global warming: the U.N. Framework Convention on Climate Change. To understand the relationship between sustainable development and the politics of the global warming treaty, however, it is necessary to first explore the extent to which both the conference and the treaty reflected the changing geopolitical context at the end of the Cold War.

Nothing underscored the atmosphere of geopolitical change surrounding the Rio Earth Summit more clearly than the rapid and unexpected global events that occurred during the third UNCED preparatory meeting, PrepCom III, held in August of 1991 in Geneva, Switzerland. On Wednesday, August 14, Russian premier Michael Gorbachev announced that the Soviet Union strongly supported the goals of UNCED and that he would personally attend the conference—a move widely perceived as a challenge to the Bush administration, which had hesitated over committing to presidential participation.[12] When the Soviet delegation to PrepCom III arrived at the conference table the following Monday, they had a strong commitment from their nation's leader and a reinvigorated role in UNCED negotiations. In what Flora Lewis of the *New York Times* called "an astonishing reversal of Soviet attitudes about international relations," the Soviets had "seized the moment on [the] global environment."[13]

When they came to work on Tuesday morning, however, the Soviet delegation did not know whether there was a Union of Soviet Socialist Republics left to represent.[14] Gorbachev had been detained and put under house arrest by a cadre of conservative party officials and military leaders, and he remained cut off from the capitol, then under a surreal, nonshooting siege. The government was in disarray, and it was unclear who, if anyone, had control in Moscow. When the dust finally began to settle during the final weeks of the year, the Soviet Union—and the Cold War—existed no longer. As Colorado senator Tim Wirth reflected dramatically in July of 1992, "The Cold War sort of stopped on one side of Rio, and something else new, different, very exciting, and certainly enormously challenging occurred on the other side."[15]

Wirth was by no means wrong, but he did not capture the whole picture either. In fact, UNCED was a conference conceived in one geopolitical paradigm and executed in another; it reflected the influences of both.

Commissioned in 1989 by the U.N. General Assembly, the U.N. Conference on Environment and Development was designed as a follow-up—twenty years later—to the U.N. Conference on the Human Environment.[16] In the three years of preparations leading up to the summit, UNCED planners developed five major objectives. First, conference planners hoped to reinforce the conclusions of the 1972 Declaration on the Human Environment with an international Earth Charter, a statement of principles built upon the 1948 U.N. Universal Declaration of Human Rights.[17] Second, as a practical guide for implementing these principles, UNCED introduced Agenda 21, a document that detailed recommendations for dealing with the endemic problems of environment and development identified at the Stockholm Conference and since then. Third, a subset of UNCED participants, including the United States and many Latin American countries, sought to create a nonbinding set of "forest principles" that would underpin a future binding convention on managing the earth's forest resources. These nonbinding principles were accompanied by two binding conventions—UNCED's fourth and fifth objectives—one a binding framework convention on biodiversity, the other a binding framework convention on climate change: the UNFCCC. Beyond these five main components, the conference would also host discussions on a spate of related (and controversial) "cross-sectoral" topics that included population growth, technology transfer, the role of women and indigenous populations in sustainable development policies, and the relationship among poverty, human settlement, and environmental degradation.[18]

From the beginning, the 1992 Rio Earth Summit shared many of the characteristics of the 1972 Stockholm Conference. For starters, the U.N. General Assembly tapped none other than Maurice Strong, the same Canadian businessman who had headed the 1972 Stockholm meeting, to serve as secretary-general for Rio. He faced many of the same frustrating environmental problems originally addressed in Stockholm—deforestation, toxic waste disposal, desertification and other land-use issues, and species extinction, among others. Perhaps more important, he also faced many of the same political obstacles.

In particular, the Stockholm Conference had left the rift between the developed and developing worlds over priorities and strategies for environmental protection largely unresolved. The Rio Earth Summit—the U.N. Conference on Environment and Development—sought to tackle the development problem head-on. As the U.S. State Department's lead man for the UNCED preparatory process, Curtis "Buff" Bohlen, put it, the "goal of UNCED is to facilitate international cooperation in more effective programs to protect our environment in a manner that is fully integrated with economic and development strategies."[19]

Not surprisingly, the integration of environmental and economic objectives meant different things to different people. Old battles over equity in international environmental regulations, over the nature and distribution of environmentally oriented development aid, and over the developed world's culpability for global environmental problems haunted the negotiations. In fact, as one U.S. delegate noted in 1991, the end of the Cold War seemed to have exacerbated tensions between the developed "global North" and the developing "global South." "Without the pressures of the old ideological divisions," Andrew Rice testified before Congress in October of that year, "the divisions between North and South are becoming more apparent." "This polarization," he argued, "is likely to continue."[20] Rhetorically, sustainable development promised to bridge the North-South divide over environmental protection. But sustainable development was a vague and flexible concept, and at UNCED the devil lived in the details. Rather than providing a solution, sustainable development became a new locus for debate.

The first and most important disagreement over sustainable development at UNCED revolved around the transfer of financial resources. As Buff Bohlen explained, when developing nations begrudgingly subscribed to the Agenda 21 recommendations for environmental protection, they did so under the assumption that the World Bank and various national and regional development agencies would provide financial assistance "over and above current levels" to meet those goals.[21] The global North—and the United States in particular—did not share this assumption, especially during a recession that reached its nadir just as preparatory negotiations for Rio reached their crux. "It is very clear," Bohlen told Congress, "that in the current economic condition within this country we cannot contemplate a big increase in foreign assistance."[22] Instead, the State Department hoped to tailor existing aid to the specific goals of environmental sustainability—a process that the Bush administration claimed would also increase the economic efficiency of development aid more generally.[23] That, according to Bohlen, was "the whole concept of sustainable development."[24]

If the overall quantity of financial resources allocated for sustainable development presented a problem for the developing world, so too did the distribution of those resources. In retooling existing aid to meet the objectives of Agenda 21, the State Department proposed a new form of oversight for the domestic economic affairs of developing nations. This presented a problem of sovereignty, as the United States and the United Nations—still dominated by the industrialized world—leveraged capital in order to establish priorities over which local populations had little say.

The Bush administration's support for the Global Environmental Facility (GEF) as the primary funding mechanism for implementing UNCED conclusions compounded this problem of sovereignty. Established in 1991 as a joint venture between the World Bank, UNEP, and the U.N. Development Programme, the GEF focused on funding programs dealing with climate change, ozone depletion, biodiversity loss, and problems of the ocean environment. The GEF continued to target the developing world, but the "global" problems it proposed to address through sustainable development practices in the developing world were, to a large extent, the problems of greatest interest to the developed world, and those for which industrialized nations bore the largest historical responsibility. Thus, under the banner of the "global," the GEF would sacrifice funding for local sustainability measures—those mandated by Agenda 21, which often had major local-level benefits—in favor of programs that ultimately served the environmental interests of the developed world. Not surprisingly, developing nations cried foul.[25]

Bohlen was not insensitive to developing-world complaints, but the Bush administration's approach to UNCED—and to sustainable development more broadly—reflected limitations imposed by a tangled web of domestic politics, American foreign policy, and international relations. U.S. participation in and contribution to the United Nations and its programs had been a topic of controversy since the founding of that organization in San Francisco in the fall of 1945. In the late 1960s and early 1970s, U.S. funding for U.N. programs, especially those that did not involve security, became a lightning-rod for criticism, especially from a subsection of American conservatives. In the late 1970s and early 1980s, the Carter administration's more cooperative "one world" approach to the United

Nations—and to international relations more generally—helped to feed this criticism, creating an even larger backlash against the United Nations and against U.S. foreign economic aid in particular.[26] Even well into the 1980s, however, the Cold War tempered this opposition. Both the United Nations and broader U.S. foreign aid were important tools in the larger effort to construct and maintain a robust global capitalist system that could withstand the affronts of Soviet-style socialism and its Eastern European and Latin American facsimiles.

Between 1988 and 1992, as the Cold War lurched into irrelevance amid a U.S.-led war in the Persian Gulf and an economic recession at home, domestic calls for a reduction in U.S. foreign aid found new credence among policymakers and the public. New funding for international environmental programs—programs that conservatives argued would slow domestic economic growth—provided grist for the mill. As Michigan congressman William Broomfield lamented, the UNCED was coming "at what I think most of us would agree is a bad time. The U.S. economy remains in recession, and worldwide economic conditions are weak.... We are being asked for more foreign aid at a time when the public clearly favors focusing on domestic issues."[27]

Thus, when Bohlen addressed Congress with his plan for implementing sustainable development through the UNCED, he stressed that "this is not another foreign aid concept."[28] In rhetoric that reflected Ronald Reagan's insistence on "trade, not aid" and consistent with the Bush party line on domestic welfare issues, Bohlen claimed that sustainable development presented a way to help the Bangladeshis feed themselves, "not through handouts, but by earning their own way."[29]

This domestic resistance to anything that looked like new foreign aid put the Bush administration in an awkward position internationally. Bush had aggressively and effectively mobilized Europe's liberal democracies as allies in the Gulf War in 1990 and 1991, and many of these allies—especially Germany and the United Kingdom—now looked to the United States to play a leadership role in environmental affairs. Japan and a number of European nations indicated that they were ready to make major financial commitments to sustainable development at the Rio Conference—the European Community proposed national targets of 0.7 percent of GDP for foreign aid—and that they would match these with deep cuts in emissions at home.[30] The Bush administration was unwilling to match these commitments; if it had, Congress likely would not have approved their budgetary components. Less than half a year away from the conference, the administration found itself isolated and on the defensive. Conference preparations continued to reflect a North-South polarization that undermined progress on the Earth Summit's five major initiatives, and the fall of the Soviet Union had removed an important geopolitical incentive to make a strong showing in Rio. In a U.S. election year, with no agreements obviously forthcoming and potentially no treaties or conventions for a head of state to sign, the increasingly unpopular American president was understandably hesitant to associate himself with an international conference on the environment that would expose him to criticism both at home and abroad. If Rio was going to fail, Bush had no good reason to tie himself to its failure.

THE UNFCCC

In the end, however, George Bush did go to Rio. As expected, a constellation of unfriendly newspaper editors, congressional Democrats, delegates from the developing world and from Europe, and spokespeople for American and international environmental NGOs pilloried the president for dragging his feet on environmental issues in favor of domestic corporate profit.[31] He was, as the *New York Times* put it, "the Darth Vader of the Rio meeting."[32] Ironically, EPA chief William Reilly, one of the prime advocates for sending Bush to Rio and a strong voice in favor of progressive environmental policy within the administration, also met a hostile reception. An embarrassing internal administration memo undercutting Reilly's earnest efforts led one sympathetic observer to characterize him as "a fig leaf for a not very committed government."[33] The administration weathered the storm. On the second to last day of the meeting, President Bush signed the convention that had, more than any other single objective at the conference, brought him to Rio in the first place: the U.N. Framework Convention on Climate Change.

Introduced in June of 1992, the UNFCCC was—and is—a nonbinding international treaty designed to stabilize the concentration of greenhouse gases in the earth's atmosphere, with the goal of "preventing dangerous anthropogenic interference with Earth's climate system," as defined by the assessments of the IPCC. Signed initially by 154 nations, the treaty relies on a Conference of the Parties (COP) consisting of delegates from each signatory nation that meets annually to oversee the implementation of the agreement's broad mandate. Though the treaty is nonbinding, each annual COP has the latitude to negotiate binding commitments, called protocols, to fulfill the convention's objectives. The original treaty assigned a permanent secretariat to serve the COPs and designated two subsidiary bodies—one for scientific and technological advice, the other for implementation—to further assist nations in meeting the terms of the convention. The UNFCCC also designated the Global Environment Fund (again, part of the World Bank), the U.N. Environment Programme, and the U.N. Development Programme as the primary financial mechanisms for implementing the treaty.[34]

The development of the UNFCCC occurred separately and in parallel to preparations for UNCED, but politically the convention on climate change was born of the larger Earth Summit. Like UNCED, the UNFCCC was built upon the contested concept of sustainable development, and debates about the function and structure of the UNFCCC reflected the same preoccupation with economic development and foreign aid.

Negotiations leading up to the introduction of the UNFCCC at Rio turned on two main issues. The first, not surprisingly, was the role of the developing world in curbing global emissions. Led in part by the host country, Brazil, developing nations feared that a climate change treaty could force poor nations to shoulder the costs of dealing with a problem, global warming, for which wealthier nations bore a disproportionate historical responsibility. Even the parsimonious Americans recognized that the global South needed financial assistance to support scientific efforts to at least monitor CO_2, and the Bush administration expressed a willingness to spend money on these general scientific efforts, specifically on projects to develop more efficient technologies through the GEF. The administration committed $250 million to the GEF, with $50 million of that sum slated for climate change.[35] But beyond this modest GEF sum (the Japanese committed $2.5 billion), the Bush administration was unwilling to provide less directed forms of foreign assistance associated with emissions reductions and sustainable development, a point of divergence from other industrialized nations. For the sake of Bush's domestic constituencies, William Reilly took pains to point out that the UNFCCC was "not a Marshall Plan for the developing countries."[36]

The developing world was not impressed. Without a healthy package of economic aid, the transition away from fossil fuels would be that much more painful. What was more, First World economies had relied on the profligate use of petroleum for their own industrialization, so this looked a lot like the industrialized world using environmental concerns to pull the ladder of development up behind them.

The end of the Cold War complicated this situation. Between the U.N. General Assembly resolution to create a convention on climate change in 1989 and that convention's introduction in 1992, several former Soviet and Soviet bloc states had begun to make the transition from centralized social-ist planning to a market economy. These were "industrialized" nations, but the legacy of Soviet rule left them with deep economic instability and some of the most dire and pressing environmental problems on the planet.[37] Latvia, Lithuania, Romania, Poland, and the new Russian Federa-tion could hardly be categorized as underdeveloped. But neither could they be expected to subsidize emissions reductions in the developing world, either through the GEF or through forms of direct foreign aid, especially in the shadow of such daunting environmental challenges at home.

Ultimately, the Intergovernmental Negotiating Committee—the body that oversaw the drafting of the UNFCCC—hammered out a tenuous compromise. The UNFCCC separated signatories into specific groups "on the basis of equity and in accordance with their common but differentiated responsibilities and respective capabilities."[38] "Annex I" countries, which included members of the Organisation for Economic Co-operation and Development (OECD) and "economies in transition" from communism to capitalism, bore a responsibility to reduce their own emissions. Of these, a subset of "Annex II" countries—mostly members of the OECD, or "devel-oped" nations—also bore a responsibility to help fund emissions reduc-tions and sustainable development in the developing world. Developing countries' responsibilities for emissions reductions, meanwhile, depended entirely on the level of financial and technical assistance they received from these Annex II nations.[39] It was a shaky and deeply problematic solution.

The second locus of dissent in negotiations over the UNFCCC involved the specific emissions targets and timetables for reducing global CO_2 and other greenhouse gases. Just as Japan and some members of the European Community called for financial commitments for emissions reductions and sustainable development abroad, so too did these nations demand that the UNFCCC set measurable and binding objectives for reducing the industrial world's greenhouse gas emissions at home. Building on the suggestions made by scientists at Villach, Bellagio, and Toronto between 1985 and 1988, the E.C. environment minister, Carlo Ripa di Meana, urged both the developed and developing worlds to devise national-level plans to reduce each nation's greenhouse gas emissions to their 1990 levels by the year 2000.[40]

As it had with committing to development aid, the Bush administration again demurred on these specific targets. The administration's position stemmed from what Bush and his advisors referred to as a "no regrets" policy on the environment. Best articulated by White House counsel Boyden Gray and the Department of Energy's David Rivkin Jr. in a 1991 issue of the magazine *Foreign Policy,* the "multiple objective steps" or "no regrets" policy approached emissions reductions as a cobenefit of addressing more concrete and tangible environmental problems like local air pollution and deforestation.[41] Gray and Rivkin characterized the state of climate science as new and highly uncertain. They ceded a few of the most scientifically conservative conclusions of the IPCC, including the existence of the greenhouse effect and the potential for warming, the rise of CO_2, and a just over 0.5°C increase in global temperature in the past century. In general, they accepted the laws of thermodynamics. But they denied that the evidence indicated a global crisis.[42] Gray and Rivkin cautioned against taking on the undue economic risks of rushed efforts to limit greenhouse gas emissions as a whole. Instead, they argued, the nations of the world should focus on more traditional environmental remediation that could hold up to a cost-benefit analysis—an analysis that would not include climate change but that, incidentally, would also reduce the overall mix of greenhouse gases. This was the low-lying fruit of climate mitigation—steps that nations would have "no regrets" taking if predictions about global warming and its impacts turned out to be wrong.

In principle, the no-regrets policy should have appealed to both the developing world and to an increasingly market-oriented environmental movement aligned with the American political left. The policy offered a holistic approach to social, economic, and environmental problems and put a positive face on potential solutions to those problems. In fact, when Al Gore argued in 1992 that the United States could reduce its greenhouse gas emissions by between 10 and 40 percent, with a net savings to the greater

U.S. economy, and that with the right leadership, the country could meet the targets proposed at Rio through purely voluntary measures, his argument spoke to the same no-regrets ethos.[43]

But the Bush administration applied its no-regrets policy within a decision-making framework that inverted one of the precepts of the environmental movement: the principle of precaution. In its simplest form—as institutionalized by the National Environmental Policy Act of 1969 and the Endangered Species Act of 1973—the environmental precautionary principle assumes that an action or policy that may be harmful to the environment ought to be considered harmful until it is proven otherwise. Under this rubric, even the most conservative reading of the first IPCC assessment—one that highlights the uncertainties of climate science and downplays the risk of climate change—would not provide grounds for inaction on the credible threat of climate change.

Bush's application of the no-regrets idea turned this approach on its head. As historian Tim Walker argues, instead of an environmental precautionary principle, the administration articulated an "economic precautionary principle," wherein any action or policy to curb environmental degradation was assumed to represent an economic threat unless proven otherwise.[44] Heavy financial commitments to sustainable development and binding emissions targets—both commitments that represented credible threats to economic growth—violated that principle.

The details of the Europeans' proposal for binding targets made it even less desirable for the Bush administration. As part of its plan to meet its greenhouse gas commitments, the E.C. Environment Ministry introduced a measure to create a simple, gradually increasing carbon tax that would raise the cost of a barrel of oil by about ten dollars by the end of the decade.[45] For the economically conservative President Bush—who had run his 1988 campaign on the back of the promise "Read my lips: no new taxes"—a carbon tax was both undesirable and politically untenable. U.S. resistance to binding targets—and to a domestic carbon tax—helped to fragment an already weakening E.C. position on the UNFCCC, and in May of 1992, the European Community voted the carbon tax down. The Dutch, the Danes, the Italians, and the Germans feared that an E.C. carbon tax not contingent upon a similar scheme in the United States and Japan would prove ineffective in curbing climate change and put the nations of Europe at a competitive economic disadvantage.[46] Similar concerns arose over the

European Community's commitments to targets and timetables more generally. German chancellor Helmut Kohl voiced his disappointment in what he saw as the failure of U.S. leadership on the climate issue. The French threatened to abandon the UNFCCC altogether.

Ironically, the trouble in the European Community only reinforced arguments from American conservatives to avoid what William Reilly called "an economic straight-jacket" of binding commitments.[47] The United States, Buff Bohlen explained, took these types of international commitments very seriously, and binding financial arrangements and emissions reductions might leave the United States holding the bag when other nations—and here he alluded condescendingly to the Europeans and the Japanese—defaulted on their responsibilities.[48] As William Reilly argued, binding commitments might prove counterproductive anyway. Surely a more toothsome UNFCCC would make many of its 154 signatories reconsider their decision to come on board. Moreover, Reilly contended, the United States could work just as well to combat climate change without targets and timetables. The Clean Air Act of 1990—one of Bush's most prized environmental achievements—had already set in motion changes that would, along with other recent conservation legislation, achieve most of the UNFCCC stipulations without limiting the flexibility of U.S. economic policy or the profitability of U.S. businesses.[49]

This last objective—maintaining the flexibility and profitability of U.S. business interests—defined the American approach to international climate change policy through the remainder of the decade. It was an approach that the United States codified by securing a commitment to "least-cost solutions" in the language of Article III of the UNFCCC itself. And it was an approach that established economic growth as the primary rubric against which to balance scientific uncertainty and environmental risk.[50]

THE TROUBLE WITH KYOTO

When William Reilly presented the text of the UNFCCC to the U.S. Senate in September of 1992, he framed it as a great success for both the Bush administration and the world at large. And as with the U.N. Conference on the Human Environment twenty years earlier, it would be a mistake to overlook the extent to which the Rio Earth Summit and the climate convention represented remarkable cooperative achievements in the protection of the global environment. In a period of massive political transition, the global political community managed to overcome significant conflicts of values and interests to tackle a major, future-tense global environmental problem head-on. The treaty was a historic one.

And yet, the UNFCCC was not actually much of an agreement, not even a compromise one. The convention had succeeded only by tabling—rather than resolving—the two most contentious problems of climate change policy. The problems of binding targets and timetables and of developing-world responsibilities haunted negotiations throughout the processes of national ratification and the first three subsequent Conferences of the Parties. Time did not heal the rifts between Americans and Europeans or between the global North and South. More than anything, it gave the parties involved a chance to regroup and dig in. Between 1992 and 1997, a changing domestic American political scene, new geopolitical wrinkles stemming from increasing economic globalization, and rapid, resource-intensive economic growth in China and India fostered a retrenchment of the adversarial positions espoused at Rio. In 1997, at the introduction of the first binding protocol of the climate convention in Kyoto, Japan, the international community recapitulated the debate over the UNFCCC. When they did, the seeds of discontent sown into the convention and fertilized by the changing political context of the 1990s began to flower into outright failure.

Adopted on December 11, 1997, the Kyoto Protocol committed the thirty-seven industrialized-world signatories of the UNFCCC to a collective 5.2 percent reduction in greenhouse gas emissions below the "baseline" levels of 1990, to be achieved between 2008 and 2012.[51] Targets for each nation differed, reflecting negotiations over differences in things like per capita emissions, GDP, and historical responsibility. The United States, for example, faced an emissions reduction of about 7 percent below 1990 levels, while Australia's target stood at 8 percent above its base level. The protocol encompassed four major greenhouse gases: CO_2, methane (CH_4), nitrogen oxides (NO_x), and sulfur hexafluoride (SF_6). It also addressed two other groups of artificial anthropogenic gases—hydrofluorocarbons and perfluorocarbons—produced primarily through the industrial processes of the developed world and used as refrigerants, lubricants, and solvents. It specifically did not include gases addressed by the Montreal

Protocol on ozone reduction. The Kyoto agreement set no binding targets for developing nations, though two of its three so-called flexibility mechanisms—the clean development mechanism (CDM) and joint implementation—involved cooperation between the developed and developing worlds in reducing developing-world emissions. A third flexibility mechanism, emissions trading, established markets in which governments and corporations could trade emissions allowances in order to reduce the economic impacts of emissions reductions in both the developed and developing worlds.

In the United States, the debates that sank the Kyoto Protocol were mainly extensions of the domestic debates over the UNFCCC. Arguments about the U.S. position on negotiating, signing, and ratifying the UNFCCC in 1991 and 1992 bled into arguments between 1993 and 1996 about the nation's responsibilities for implementing that treaty, which in turn informed arguments over the U.S. position on negotiating, signing, and ratifying the Kyoto Protocol. Again, disagreements over the U.S. position on targets and timetables and on developing-world responsibilities played out in the language of scientific uncertainty and economic risk.

Backed by the new Clinton administration, as early as 1993 select Democrats and environmentalists began to push for major revisions to President Bush's fiscally conservative no-regrets plan for climate change, which they complained was merely "a compendium of programs and actions that are being undertaken for other reasons."[52] Vice President Al Gore and Undersecretary of State for Global Affairs Tim Wirth continued to support reducing U.S. emissions (for example, through transportation infrastructure and energy and fuel efficiency), and they urged an aggressive leadership role in international climate change negotiations. The cost savings of energy efficiency and the new green technology jobs that would result from aggressive U.S. emissions reductions, they argued, would lead to economic growth at home and would put the nation in an economically advantageous position abroad.[53] On a broader scale, targets and timetables would provide the market system with the certainty it required to adapt to new regimes of energy use and emissions controls.[54]

Congressional Republicans, backed by industry representatives like Michael Barrody of the Global Climate Coalition (whose members included British Petroleum, Shell, DuPont, Ford, and General Motors, among others), offered a multifaceted response that built upon the twin pillars of scientific uncertainty and economic precaution. First, according to political conservatives' reading of the IPCC first assessment, the science was not yet "in." Conflicting theories suggested that climate change may not be that big of a problem if the rise in global temperatures occurred at the low end of the predicted range (which conservatives suggested was likely); that climate variation might simply result from "natural" environmental processes, not anthropogenic increases in greenhouse gas concentrations; and that the global climate system might have a self-correcting mechanism that would counteract the more extreme cases of greenhouse warming. The IPCC had addressed and largely dismissed each of these issues in turn; but IPCC recommendations notwithstanding, many conservatives argued that in reality scientists just did not know.[55]

Second, opponents of strong domestic action on global warming argued that because of these scientific uncertainties, climate change did not merit what they characterized as an economically disastrous crash course to reduce emissions. In very testy congressional testimony in 1992, for example, a combative Senator Mitch McConnell from Kentucky predicted that the fight against global warming, "the most expensive attack on jobs and the economy in this country," would cost America more than a million jobs.[56] This argument extended to the international arena, where binding targets and timetables would limit the flexibility of U.S. foreign policy and compromise the American business community's international transactions in the face of growing Chinese and Indian economic power. The inauguration of a Democratic president in 1993 temporarily put the Republicans on the defensive, but for the most part the terms of the debate over an international treaty on global warming remained the same in the mid-1990s.

The Kyoto Protocol did introduce a few new elements, however. The first was a foreign policy wrinkle, another artifact of the end of the Cold War. Bill Clinton had won the presidency in part by making domestic social and economic promises in a new era of American foreign policymaking. With the fall of the Soviet Union and the end of the first Iraq war, the central objectives of twentieth-century American diplomacy—promoting liberal democracy, protecting U.S. markets abroad, and stemming the tide of Soviet influence and communist expansion—seemed either irrelevant or mostly accomplished.[57] As a result, Clinton's first administration proposed mostly limited foreign policy initiatives that built upon his domestic policies and upon the liberal approaches espoused by his

Democratic predecessors, including increased food and development aid. The administration sought to use the traditional organ of the State Department—and its quasi-independent subagencies, the U.S. Agency for International Development (USAID), the U.S. Information Agency (USIA), and the U.S. Arms Control and Disarmament Agency (USACDA)—to promulgate Clinton's foreign aid policies. Conservative members of Congress, however, objected to unnecessary spending that threatened the dual conservative precepts of limited government and free markets. After the Republicans gained control of both houses of Congress in the 1994 midterm elections, conservatives found that they had the power to push their conservative principles.[58]

Constitutionally, Congress itself does not have the power to make foreign policy; rather, its directive is to provide advice and give consent to the president on presidential initiatives. What Congress does have direct control over is the budget of the State Department. Coupled with a broad interpretation of "advise and consent," this budgetary oversight has enabled Congress to massage its advisory role into real power, and select legislators have consequently exercised significant influence over American foreign policy throughout the nation's history.[59]

As chair of the Senate Foreign Relations Committee beginning in 1995, North Carolina senator Jesse Helms took his role as a foreign policy masseur seriously. He flatly refused to submit a State Department budget to the Office of Management and Budget for approval until the State Department had been severely reduced in size, its policymaking autonomy curtailed. Among other things, Helms demanded that the administration excise the agencies that traditionally carried out aid-oriented policy, particularly USAID. The result was a two-year standoff during which the State Department was granted only enough emergency funding to carry out its most basic diplomatic functions, until it trimmed enough fat to satisfy Helms's committee. For Helms, one piece ripe for the trimming was Secretary of State Warren Christopher's new focus on promoting sustainable development.[60] The standoff lasted until 1996—less than a year before Kyoto negotiations began—and the episode set the tone for climate mitigation debates that themselves would largely revolve around development.[61]

It was within this context of foreign policy limbo that the Senate used its "advise and consent" mandate to create another wrinkle in the politics of the Kyoto Protocol. Led by Republican Chuck Hagel of Nebraska and

Democrat Robert Byrd of coal-rich West Virginia, in July of 1997 the Senate put its opposition to a binding protocol under the UNFCCC in writing. In the Byrd-Hagel Resolution, the Senate stated that it intended to ratify no treaty that would either (a) "mandate new commitments to limit or reduce greenhouse gas emissions" without "new specific schedules to limit or reduce greenhouse gas emissions for Developing Country Parties" or (b) "result in serious harm to the economy of the United States."[62] The resolution also stipulated that the Senate would not consider any treaty that was not accompanied by a detailed plan of regulatory or legislative action to meet its commitments, as well as an analysis of the costs of those actions.

If the terms of the climate change debate remained the same throughout the 1990s, the Byrd-Hagel Resolution underscored the extent to which the political calculus had changed between 1992 and 1997. In 1992, Bush had somewhat begrudgingly signed a weakened treaty that did not assign responsibility for emissions reductions to the developing world, and when he did he met serious criticism for his foot dragging and lack of leadership from environmentalists and congressional Democrats. But in the mid-1990s, traditionally Republican business and industry lobbyists had begun to court traditionally Democratic labor organizations, notably the automobile workers' union (the UAW) and the United Mine Workers of America, to an antitreaty position.[63] This coalition lobbied against the treaty intensely in the years and months leading up to Kyoto. They targeted Congress specifically, driving a wedge between the administration and the Senate on the issue. By 1997, fueled by the dual rhetoric of scientific uncertainty and economic precaution (the latter cast in terms of American jobs), opposition to a robust climate treaty had grown to such a degree that not a single senator was willing to support a binding commitment to emissions reductions limited to industrialized nations. Byrd-Hagel passed 95–0.

Both the State Department stalemate and the Byrd-Hagel Resolution put Undersecretary of State Stuart Eizenstat and the rest of the Clinton administration's Kyoto negotiating team in something of a pickle. In light of the more definitive conclusions of the IPCC's second assessment report in 1995, the United States had agreed to the Berlin Mandate at the first Conference of the Parties that same year. This mandate committed the developed world to set binding targets on emissions reductions by as early as 1997. But these commitments violated both the economic precautionary principle of "no regrets" and the flexibility in international economic affairs that "no regrets" had been mobilized to protect. Pushed by the European Union, the Berlin Mandate also specifically exempted developing-world parties from the 1997 targets. Byrd-Hagel made it clear that the Senate would reject a treaty based on the Berlin Mandate. This left the United States and the European Union struggling to bring the developing world on board with binding commitments, but here again congressional politics tied negotiators' hands. The State Department fiasco had demonstrated that any plan for significant new development aid would fail in the Senate, robbing the U.S. negotiating team of its one major incentive—a financial one—for luring the developing world into some form of commitment. Meanwhile, the American public (with some coaching from congressional conservatives and antitreaty groups like the Global Climate Coalition and the Global Climate Information Project) had all but rejected anything smacking of a "carbon tax," and the newly signed North American Free Trade Agreement (NAFTA) severely limited other legal and regulatory mechanisms for enforcing emissions reductions.[64] For Eizenstat, things did not look good.

Ironically, it was from this impossible position that the U.S. negotiating team helped to devise the flexibility mechanisms that earned the Kyoto Protocol the international support it eventually enjoyed. The first of these, called joint implementation, allowed Annex I countries to meet emissions reductions standards by investing in greenhouse gas reductions in other Annex I countries (typically the "economies in transition" eventually listed in Annex B of the Kyoto Protocol). In part, this was a response to the European Union's demands that Europe be allowed to reduce its emissions as a collective unit. Perhaps more important, joint implementation provided a sort of global no-regrets framework that gave all nations the incentive to address least-cost options for emissions reductions.

The second flexibility mechanism, the clean development mechanism, provided for a similar least-cost approach to reducing developing-world emissions. Under the CDM, Annex I countries could meet their domestic targets by investing in emissions reductions and sustainable development programs in developing-world nations that had no specific responsibilities under the Kyoto Protocol. Much of this investment would come in the form of "technology transfer," whereby the industrialized world would replace inefficient technologies and industrial processes with new and more efficient methods and machines. Like joint implementation, the CDM gave

Annex I nations the flexibility to take a least-cost, global approach to emissions reductions. It also served as an incentive for developing-world countries to cooperate with Annex I countries' emissions reductions objectives, by providing the developing world with certain environmentally specific forms of development aid.[65]

By far the most important and controversial of the flexibility mechanisms built into the Kyoto Protocol was emissions trading, commonly known as cap and trade. Under an emissions trading system, a central organization—in this case the United Nations, under the Kyoto Protocol—sets a limit (a cap) on the total amount of a pollutant that can be emitted. The organization then allocates standardized credits that give their holders the right to emit a certain amount of the controlled pollutant. Those credits—carbon credits in this case, representing greenhouse gas emissions equivalent to a fixed amount of CO_2—can then be bought and sold as a commodity in national, regional, and global markets. A company that holds carbon credits has the option to either reduce its own emissions and sell the unused credits or to emit at a certain level and use its own credits, depending on the cost of emissions reductions and the market price of emissions permits. Under the Kyoto Protocol, these credits remain valid for a fixed period, such that an overall reduction in the emissions "cap" to below 1990 emissions levels can be phased in over time.[66]

Proposed by Eizenstat and the U.S. negotiating team, cap and trade was a late addition to the Kyoto negotiations that clearly reflected American interests. The Clinton administration modeled the system after a similar program that curbed acid rain by limiting sulfur dioxide emissions. The Grand Policy Experiment, as its proponents dubbed it, was the cornerstone of the 1990 Clean Air Act passed under the Bush administration, and by the mid-1990s it had exceeded expectations as a cost-effective, market-based solution to pollution control.[67] Backed by Vice President Al Gore (formerly a strong proponent of a carbon tax), cap and trade sidestepped the political quagmire of taxation while also satisfying the second and third stipulations of the Byrd-Hagel Resolution—that any binding commitment must not harm the American economy and that it must include a detailed implementation plan with a thorough cost-benefit analysis. Just as important, cap and trade provided a flexible, market-based alternative to European proposals for legal mandates and direct regulatory intervention, steps that conservatives and free-market economists, with a pejorative nod to the failed Soviet-style command economy, called the "command-and-control" approach.[68]

In the fall of 1997, the Europeans wanted no part of the American proposal. What Eizenstat articulated at Kyoto was a compliance market rooted in a commodified right to pollute. Although some U.S. environmental organizations expressed philosophical discomfort with the system, in general Americans had become comfortable with market-based approaches that modified property rights regimes to deal with environmental problems. Sulfur dioxide was the most recent and successful example, but similar programs existed for water allocation and for managing quotas in declining fisheries.[69] Europeans, especially those who had supported the 1992 Earth Charter based on the original U.N. Universal Declaration of Human Rights, found the scheme troubling at best. The right to pollute was anathema to the European Union's spirit of collective environmental responsibility, and European environmentalists distrusted both markets and the Americans who proposed using them to mitigate climate change. It was not until Gore actually flew to Kyoto in December of 1997—where he explained, among other things, that cap and trade would facilitate compromise on the deeper emissions cuts that the European Union had been pushing for—that Eizenstat was able to get the Europeans to agree to emissions trading.

Eizenstat's sincere efforts to broker a compromise between the European position and the strictures of Byrd-Hagel notwithstanding, in the United States the Kyoto Protocol was dead on arrival. The Senate had no intention of softening its position on developing-world participation, and business and industry coalitions continued to highlight the economic costs of addressing a problem supposedly still rife with scientific uncertainty. Cap and trade promised to provide a nontax mechanism for greenhouse gas reductions, and many American companies eventually made money trading carbon credits; but committing to Kyoto's binding targets still threatened to limit America's foreign policy options and restrict Americans' flexibility in the global economy. Most important, despite joint implementation, the CDM, and emissions trading, behind the Kyoto Protocol lay recognition that the excesses of the petroleum-based capitalist economy must be curtailed not just abroad but also at home. Already under siege from conservatives on a number of domestic issues, the Clinton administration decided to cut its losses. The president never sent the Kyoto

Protocol to the Senate for a vote. In the boom of the late 1990s, Americans were unready and unwilling to sign an agreement ultimately aimed at changing the lifestyle they had worked for and defended for so long.

A NEW RUBRIC?

At first glance, the stories of the UNFCCC and the Kyoto Protocol seem to depart from the stories of science politics that dominate the history of global warming. The failure of the Kyoto Protocol in the United States was certainly not a failure of the scientific community. American resistance to a binding agreement turned on a resurgence of domestic fiscal conservatism, changes in the geopolitical landscape at the end of the Cold War, and a long-standing tradition of prioritizing economic interests above all else in American foreign policy decision making. In fact, for the scientists who had used the "forcing function of knowledge" to influence policy by speaking truth to power since the 1970s, the treaty-protocol process initially seemed to validate their science-first, top-down approach. With the UNFCCC and later Kyoto, it seemed, power had begun to listen.

But the politics of international governance that sunk the Kyoto Protocol grew directly out of the science politics that had earlier weakened the IPCC and undercut the nascent UNFCCC. In the 1980s and early 1990s, opponents of serious national and international efforts to curb emissions realized that the first, best way to influence the political discussion about global warming was to challenge the science behind it. By weakening scientific consensus and challenging scientific certainty, representatives from business and industry sought to insulate themselves from blame—and regulation—in future political debates that might unfold in legal and moral terms. If the IPCC represented the front end of this effort—a forum for science as a form of preemptive politics—the UNFCCC and Kyoto Protocol saw this strategy bear fruit. The no-regrets policy, based on economic precaution and questioning scientific certainty, was tenable only as long as the IPCC assessments continued to reflect a conservative view of the science behind global warming. Scientific uncertainty justified greater economic precaution. Advocates of greater economic precaution consequently highlighted scientific uncertainty. This was the structure of science-first opposition shaped by science-first global warming advocacy.

Moreover, the turn toward economics as an arbiter of climate change policy was not a turn away from a deep faith in the power of science to define rational policy on global warming, but rather an extension of it. In the 1990s, economics became the lingua franca of climate change policy. Just as scientists had expected the forcing function of knowledge to force rational policymakers' hands toward climate mitigation plans in the 1970s, so too did environmentalists, industrial leaders, politicians, and diplomats suggest that their particular hard-nosed economic analyses would dictate a rational and obvious course of action in the 1990s. Even more so than in debates about the science of global warming, in debates about the economics of global warming, advocates on all sides relied on numbers to determine winners and losers. Conflicting economic scenarios like those propounded by Al Gore, who argued that climate mitigation would save money and create jobs, and Mitch McConnell, who argued that climate mitigation would ruin the U.S. economy and cost jobs, perhaps did a poorer job than scientific debates at hiding their political motives. But ultimately, the language of science and the language of economics masked a constellation of moral and political concerns in artificially rational, objective terms.

It may be uncharitable to the United States to neglect the intransigence of developing-world emitters like China and India in stalling negotiations at the Kyoto conference, but ultimately the failure of the Kyoto Protocol was an American one. Most obvious, despite its responsibility for a quarter of the world's emissions at the time, the United States was the only industrialized nation that failed to ratify the agreement, and this after months of negotiations meant to tailor the protocol to U.S. specifications. Perhaps more important, however, both the Clinton administration's goals and the Senate's resistance stemmed from the characteristically American effort to externalize both the problem of global warming and its potential solutions. Americans' focus on the role of the developing world reflected a desire to deal with the problem primarily there, in the developing world, rather than in Pittsburgh or Dallas or Detroit. Kyoto's flexibility mechanisms ensured that Americans could meet their commitments to the agreement without any real, dramatic changes to U.S. modes of production or lifestyles of consumption.

This was a new twist on the problematic primacy of science in global warming politics, in a new post–Cold War context. Scientists worked tirelessly to communicate the global nature of the sources, mechanisms, and impacts of CO_2 and global warming, and they codified this global understanding in the IPCC. But the primacy of science in global warming politics meant that scientists' descriptions of the problem also defined the scales at which the global community sought to solve the problem. From a global scientific perspective, reductions in CO_2 emissions from India or China or Japan provided the same benefit as equal reductions from the United States. In an interconnected global world, economic calculations of least cost could supplant the difficult discussions about development priorities, social equity, and historical responsibility that came with geographically specific emissions targets.

Americans' predilection for economic analysis as a guide to policy—especially foreign policy—was by itself nothing new in the 1980s and 1990s. In his classic *Tragedy of American Diplomacy*, for example, William Appleman Williams demonstrated the extent to which policies of economic expansion abroad designed to support economic prosperity at home have guided the U.S. approach to diplomacy since the 1890s. But the triumph of global capitalism at the end of the Cold War helped to establish the unabashed primacy of economics in international climate negotiations unique to the end of the twentieth century. The debate over the Kyoto Protocol in the United States was less a debate over whether this economics-first approach was the right one than it was a debate over what the right economics-first approach would look like. For Stuart Eizenstat and the Clinton administration, the flexibility mechanisms built into the Kyoto Protocol—joint implementation, the CDM, and emissions trading—made it possible to use foreign investment and development aid to insulate Americans against potential economic shocks resulting from domestic climate mitigation. As proponents of cap and trade argued, the new system would actually create new markets for American investors. For anti-Kyoto members of Congress and their allies in business, industry, and organized labor, however, the threat of competition from developing-world manufacturers, unfettered by the strictures of a greenhouse gas regime, threatened to overwhelm these potential opportunities. It was a case of new markets against free markets. In the Senate, free markets won. Meanwhile, on the Keeling Curve, CO_2 continued to rise.

EPILOGUE

Climbing Out from Behind the Curve

LOOK AGAIN AT THE KEELING CURVE. BY NOW, THE IMAGE SHOULD BE familiar, not just as a measure of CO_2, but as a narrative thread through the history of global warming. Imagine the undulating line that connects the monthly average of atmospheric CO_2 weaving through a patchwork tapestry of related and overlapping historical contexts. Here it is sewn into the optimistic science politics of the early Cold War; there, in a loose and awkward stitch, it joins Cold War fears, opposition to the conflict in Vietnam, and a burgeoning American environmental movement. Each new piece of fabric or color of thread makes the curve look slightly different. Each new context requires a new interpretation of atmospheric CO_2, and, taken together, these reinterpretations constitute the political history of global warming.

The key to understanding this history of global warming is to think about the iterative stories of CO_2, first in their specific historical contexts and then in their aggregate; that is, to return to the Keeling Curve, to look first at the annual oscillations in isolation and then as a larger trend. But unlike in the curve, in the narrative the specific historical contexts become more difficult to interpret, more subject to rapid and dramatic change, as they approach the present. The lens of history is valuable in part because time tends to smooth out the noisy extremes of any particular moment. To use an analogy from statistics, time does qualitatively for historical evidence what regression does quantitatively for numerical data: it reveals the patterns and evens out the bumps. Meanwhile, the larger interpretive framework of tragedy helps make sense of the iterative stories of climate change only insofar as it gives them a recognizable beginning, middle, and end. The end is necessarily artificial—life goes on—but it serves an important purpose in that it defines the failure of the past and marks a point from which to move forward. An attempt to tell the comprehensive, up-to-the-minute story of global warming politics runs the risk of turning into either a painful protracted Act V in the style of *King Lear* or, worse, a sort of climatic *Tristram Shandy,* a farcical collection of episodes that would do more to hide the structures of the story than reveal them.

Then again, a good deal has happened in global warming politics since the turn of the twenty-first century, much of it supremely relevant to the larger themes of global warming history. In 2003, for example, CO_2 entered the American legal system in a new way when twelve states, supported by environmental organizations, sued the Environmental Protection Agency to regulate CO_2 and other greenhouse gases as pollutants. Meanwhile, the Kyoto Protocol, which the United States never ratified, entered into force in 2005, and its myriad shortcomings have informed renegotiations of and new ideas about the U.N. Framework Convention on Climate Change and its processes. The intricacies of the 2009 Conference of the Parties in Copenhagen could sustain a full monograph on their own. Meanwhile, the profile and credibility of the Intergovernmental Panel on Climate Change has waxed and waned. In 2007, Al Gore and the IPCC won what many considered to be a politically motivated Nobel Peace Prize for their work on climate change, but the award coincided with the beginning of new attacks on the IPCC's consensus-making process that culminated in a scandal involving the perceived manipulation of climate data by IPCC contributors at the University of East Anglia in 2009. And in 2013 President Barack Obama made headlines by framing climate change policy as what the press likes to call a "legacy issue" during his second term. How can history help us better understand these developments in climate change politics, and how can it help us interpret new developments in the future?

Two important developments in climate change discourse since 2000 reveal how history can be useful for understanding the changing world of global warming politics. Perhaps the most novel recent development came in the realm of law, in the Supreme Court case *Massachusetts v. EPA*. In some ways, the case is a coda to the Kyoto Protocol. In 1999, nineteen environmental organizations filed a petition with the EPA, asking the agency to regulate CO_2 as a pollutant under the regime regulating vehicle emissions outlined in the Clean Air Act.[1] The petitioners sought to use the existing machinery of domestic environmental regulation to accomplish some of the greenhouse gas reductions that the Kyoto Protocol would have stipulated, had the Senate ratified it. In that sense, the EPA petition was a climatic plan B for the new millennium.

The case began to take shape in 2003, when the George W. Bush administration EPA officially rejected the petition.[2] CO_2, the EPA claimed, did not constitute a "pollutant" under the Clean Air Act, and the agency lacked the authority to regulate CO_2 and other greenhouse gases outside of that act's pollution mandate. In 2005, represented by James Milkey of the Massachusetts Attorney General's Office, twelve states joined the original petitioners in an appeal to the District of Columbia Circuit Court.[3] The court of appeals upheld the EPA's decision, but within a year the case found its way to the Supreme Court.

At the Supreme Court level, the case revolved around three main issues. The first was whether any or all of the petitioners had a right to claim injury from the nonregulation of greenhouse gases—that is, the procedural issue of standing. The majority of the court decided that only one of the many plaintiffs had to meet the requirements for standing and that Massachusetts's vulnerability to rising sea levels as a result of global warming met that standard. Once the court granted Massachusetts standing, it was left to decide, first, whether CO_2 and other greenhouse gases met the standards of pollution under the statute that allowed the EPA to regulate emissions (the Clean Air Act); and second, if CO_2 did qualify as a pollutant, whether EPA administrators had the discretionary authority *not* to regulate emissions in light of scientific uncertainty—that is, whether scientific uncertainty was a valid reason for the agency to sit on its hands. The court ruled in favor of the plaintiffs on both of these issues as well and asked the EPA to reconsider whether it had a valid reason not to regulate greenhouse gases.[4]

At first, *Massachusetts v. EPA* looks like a significant break from the structures of discourse and advocacy that dominated the history of global warming in the twentieth century. Led by environmental organizations now committed to a cause that they had eyed warily for a long time, the plaintiffs in the case inserted CO_2 into a new legal context—a context that had mechanisms for interpreting science but in which questions of law, rather than questions about modeling and uncertainty, took precedence. For the court, the science had already largely been settled. Building on a "strong consensus among qualified experts," the justices noted, "the Government's own objective assessment of the relevant science" had already provided sufficient evidence that global warming would have significant negative impacts on human and natural systems.[5] Rather than digging into the science, the court focused on the statutory responsibilities of the EPA. In fact, the third part of the court's decision—that scientific uncertainty did not constitute a valid reason for nonregulation of greenhouse gas emissions—specifically reasserted the primacy of the Clean Air Act's statutes over the scientific conservatism and deference to economic precaution that had characterized the EPA's initial response to the petition.

If the legal venue and the focus on statutory language represented new wrinkles in the political history of global warming, however, the *Massachusetts v. EPA* story also contained much that was familiar. First, the entire case represented yet another attempt to shoehorn the intractable problem of climate change into the existing structures of American environmentalism. The breadth and scope of the Clean Air Act enabled Milkey to frame CO_2 as a pollutant like any other; but even as it handed down its favorable decision, the court recognized that in their scope and causal complexity, greenhouse gases presented a unique regulatory challenge within the pollution paradigm. Scientists certainly knew more about global warming when *Massachusetts v. EPA* began to take shape in 2003 than they did in 1963, when Keeling published his article about CO_2 as pollution; as a society we have taken a much greater political interest in the subject over the past half century. But as the court case demonstrates, the basic intractability of the problem remains.

The outcome of *Massachusetts v. EPA* also recapitulates a familiar story. Milkey and his fellow plaintiffs hoped to use the statute for regulating vehicle emissions under the Clean Air Act to force the EPA to regulate CO_2 and other greenhouse gases more broadly. But beyond regulations for miles per gallon in new vehicles, the plaintiffs presented no obvious mechanism for this regulation. They assumed that given the authority and the mandate to regulate greenhouse gases (or to allow individual states to do so on their own), the EPA would develop appropriate emissions standards based on its assessment of the best available science. The Supreme Court noted that the EPA could not hope to reverse global warming, only to regulate greenhouse gases to help slow global warming. But when the justices remanded the case to the EPA, they left it to that agency to decide what steps it should take to slow global warming, and by how much. Again, these decisions would be based on expert scientific opinion both from within the agency and from the broader scientific community. The court thus mandated that the EPA use good science to make good policy.[6] It was the law forcing the forcing function of knowledge. Even in a new legal context, science continued to reign supreme in global warming discourse.

If *Massachusetts v. EPA* represented a rhetorical victory for global warming advocates, the case had far less public impact in the United States than a concurrent development in global warming politics: the release of Al Gore's 2006 documentary, *An Inconvenient Truth*. The film both recapitulates much that was old in climate change discourse and introduces something new. Alongside the court case, it demonstrates how the history of global warming can help us interpret the issue's contemporary iterations.

Adapted from a slideshow and talk that Gore began to give after his defeat in the controversial presidential election of 2000, *An Inconvenient Truth* offers a protracted scientific exposé on the causes and impacts of global warming. Capitalizing on his political celebrity and weaving engaging autobiographical anecdotes, humor, and a simple message about personal morality into a synthesis of up-to-date scientific research, Gore pitched *An Inconvenient Truth* at exactly the right level for an American public familiar with but not particularly knowledgeable about global warming. The film debuted in May 2006 and quickly became a critical success. At the box office, it grossed $24 million domestically and $26 million abroad. It won two Academy Awards. Perhaps more importantly, polling showed that the movie screened well even in conservative suburban markets.[7] *An Inconvenient Truth* did not just make money; it inserted CO_2 and global warming into the consciousness of the American middle class, a group that climate scientists had largely ignored until the 1990s. This was something new.

There was plenty that was not new, however. The familiarity of the story of *An Inconvenient Truth* began with Al Gore. Much of Gore's success at presenting a reasonably accurate synthesis of climate science in the film came from his longtime interest in global warming as both a scientific and a political issue and from his relationships with the scientists who had brought the issue before Congress in the early 1980s. The film also carried a political subtext reminiscent of those early hearings. In the early 1980s, Gore and his fellow congressional Democrats had used global warming to attack Ronald Reagan's policies on energy, defense, and the environment. In 2006, Democrats had a similar predilection for highlighting George W. Bush's failures. With global warming already well established as an important political issue, Gore made sure that his audience recognized the contrast between his positions and those of the candidate who had beat him in 2000. The opening line of the slideshow set the tone. "I am Al Gore," he said, "I used to be the next president of the United States."[8]

The film's main approach to global warming was also familiar. Like so much in the discourse on climate change, Gore's movie focuses squarely on validating, interpreting, and communicating science. Gore takes great pains to explain the basics of the greenhouse effect, the earth's energy budget, and rising CO_2. He also goes out of his way to defend the scientific consensus that the problem is real, caused by humans, and potentially catastrophic. He uses Naomi Oreskes's 2004 survey of nearly a thousand peer-reviewed articles on climate change to demonstrate that the science of global warming has been settled.[9] If nothing else, *An Inconvenient Truth* is supposed to be a lesson in that science and in what that science means for ecosystems, species, and humans.

And yet, even with the science "settled," Gore gave little that is new in the way of "next steps" on global warming, beyond individual action and a vaguely defined "political will." Solutions, Gore argues in the movie, already exist, first in the form of personal behavior and second in the form of the international framework for CO_2 mitigation laid out in the Kyoto Protocol, which the United States still had not ratified. During the movie's credits, Gore presents ways that individuals might fight global warming: recycle, drive a hybrid, conserve energy, and other familiar "green" behaviors. One of the most prominent suggestions is to "encourage everyone you know to watch this movie." If there was something different about Gore's suggestions it was that he put global warming solutions in the middle-class consumerist terms of the broader environmental movement and spoke to a wide public audience. But even in a new popular context, the focus on knowledge and its power to influence politics remained the same.

The familiar commitment to science as both catalyst for an arbiter of global warming solutions, common to both the *Massachusetts v. EPA* case and *An Inconvenient Truth*, speaks to the persistence of the most important feature of global warming politics since the 1950s: the primacy of science. The nature of the problem—global, causally complex, long-term, diffuse—has privileged science as the dominant language for dealing with global warming since its early iterations in the late 1950s, at the height of the Cold War. Scientists continued to drive global warming advocacy throughout the twentieth century. Though the political context in which they developed that advocacy changed dramatically, most of the major economic, legal, and scientific mechanisms that they helped develop to deal with global warming revolved around the production and validation of knowledge. It is telling that the Nobel Committee awarded the 2007 Nobel Peace Prize to Al Gore, for his work publicizing the issue of global warming, and to the IPCC, for its role in articulating the scientific consensus around which advocates hoped to build international climate change policy.

Winning a Nobel Peace Prize for well-intentioned hard work on global warming hardly makes the stuff of tragedy. And yet, as the Keeling Curve passes the 400 ppm mark, it is difficult to ignore that the lauded science-first approach to global warming has thus far failed to meaningfully affect the curve itself.

The tragedy of the history of global warming is that the primacy of science has tended to limit the nature and impact of global warming advocacy by privileging science as the best, and often the only, way to deal with the problem. In fact, the science-first approach has at times actually undermined the kinds of moral and political discussions that many global warming advocates, ironically, have relied on science to foster. The results are tragic in two senses. First, the story fits the narrative mold of a tragedy in that the protagonists' means of accomplishing their goals in the end undermine those goals. But second, and more profoundly, the failure to effect meaningful change on global warming is tragic in the colloquial sense, with real and sometimes devastating consequences. Ask an Inuit who has watched her traditional way of life disappear with the Arctic sea ice about climate change or a Pacific Islander who has watched his island sink into the ocean. They will tell you about this aspect of the tragedy.[10] As we have continued to lodge moral and political debates in the language of scientific consensus and uncertainty, those debates have delayed strategies for mitigating and adapting to global warming at the expense of vulnerable organisms, ecosystems, and species—including humans.

It is easy to go too far with this line of argument. It would be wrong to conclude that we should stop working toward global-scale political strategies to address global warming or that scientists should cease to play a primary role in building these frameworks for change. The primacy of science in global warming politics did not cause CO_2 to rise, nor does the history of global warming support the claim that science or scientists have actively prevented a decline in the slope of the Keeling Curve. If anything, the efforts of scientists, environmentalists, and select politicians, diplomats, and corporations to use science to fight global warming have probably tempered the increase in atmospheric CO_2 over the last two decades. Institutions like the IPCC and treaties like the UNFCCC provide both guidance and political justification for the national-level efforts essential to curbing emissions worldwide. Science helps to give these international agreements their authority; and at this macro level, science and its numerical language continue to help define the problem and measure our collective progress— or lack of progress—in solving it.

But it is also important to understand that information is not the same as political change. One of the reasons that we are behind the curve on global warming is that historically we have conflated these two things. Scientists, environmentalists, and international leaders have helped create and have participated in international organizations designed to make and apply knowledge. In thinking about the problems described in the models and curves of climate science, we have tended to turn to those same models and curves for solutions. We think of global warming policy in the form of international treaties like the UNFCCC, based on a global vision of climate science codified by the IPCC. But these institutions lack the political power to meaningfully affect the production and consumption of energy—the basic cause of rising CO_2—at its most important local, regional, and national scales. Those things require local-, regional-, and national-level social and political changes that often have very little to do with the forms of knowledge that climate science can provide.

In *Why We Disagree about Climate Change*, Mike Hulme introduces several approaches to the question of why the world has failed to address climate change in a meaningful way. Of these, the most convincing is Hulme's discussion of the nature of the problem itself. Climate change is what some political scientists refer to as a "wicked problem." As Hulme explains, wicked problems tend to defy clear description, to involve complex interdependent systems, to require decision making based on incomplete or contradictory information, to occur in an undefined time frame, and to lack a clear rubric for evaluating solutions (of which there are often none).[11] Climate change is a category of problem that includes most of the intractable challenges of the twenty-first century: poverty, crime, drugs, and terror, to name a few of the most obvious.[12] Though climate change has not yet received this honor, I like to joke that perhaps the best way to identify a wicked problem is to ask whether the federal government has declared war on it.

Articulating the intractability of the problem is important in approaching the history of climate change, but "wickedness" is only part of the story. The way in which we as a society have responded to other wicked problems differs markedly from our collective approach to global warming. That may be changing, however. A final and, I think, hopeful development in the political history of global warming from the 2000s underscores what has been missing in the top-down, science-first approach that has dominated global warming advocacy in the past.

In the mid-1990s, even as Al Gore and Stuart Eizenstat were making headlines trying to turn the UNFCCC into an effective and politically acceptable Kyoto Protocol, a much quieter form of local and regional climate change advocacy began building in U.S. towns, cities, and states. It was a movement led by mayors, governors, and city councils to create "climate action plans," locally and regionally specific strategies for mitigating greenhouse gas emissions and adapting to global warming impacts that enable municipalities and states to take responsibility for their emissions in the absence of federal climate change policy.[13]

Initially, climate action plans looked like a new twist on a familiar science-first story. In the 1980s and early 1990s, supported in part by grants from the EPA, states and municipalities began to make emissions inventories in order to study their local and regional greenhouse gas contributions. Organizations like Cities for Climate Protection and the International Council on Local Environmental Initiatives provided resources for cities to study emissions, mitigation potential, and climatic vulnerability, in large part to build local- and state-level support for nation and international climate change policy.[14]

These communities' responses to the Byrd-Hagel Resolution and the failure of the Kyoto Protocol took a turn away from science-first, however, revealing something different about the meaning of CO_2 in these local and regional contexts. Rather than following a wait-and-see approach, or deferring to national and international policies to deal with what scientists described as a global problem, in the early part of the decade communities began to plan and execute local emissions reductions programs that would comply with Kyoto Protocol targets. The result has been "bottom-up" or "multilevel" governance, built as much upon communities' collective sense of moral responsibility as on the science-to-policy models that have driven global warming advocacy in the past.

Part of the impetus to act locally is still to force national-level policy by demonstrating broad-based support and providing examples of how local-level emissions reductions work. But part of the impetus to deal with global warming locally also stems from a real desire to build better, healthier, and more responsible communities. The approach is similar to how communities have historically dealt with other "wicked" problems like poverty, crime, and drugs. Climate action plans tend to look to Kyoto for their ultimate emissions objectives, but in their nuts and bolts they have much more in common with (and some overlap with) zoning regulations, municipal parks plans, state and federal unemployment programs, public health initiatives, and even Neighborhood Watch associations. They are in some ways a manifestation of the oft-repeated admonition to "think globally, act locally." But rather than focusing on the poles of that dichotomy—local, typically personal action on the one hand for the sake of collective global goals on the other—climate action plans have begun to use the mechanisms of municipal, county, and state governance to shoot for the middle. In weaving climate mitigation into the structures of daily life, climate action plans give communities a chance to reorganize themselves as if the abstract global climatic good mattered to everyone, every day. This is something new.

Global warming is not a problem that can be solved in any traditional sense of the word. In fact, by at least one measure, we have already failed to solve it. The IPCC estimates that we are "locked in" to at least 1–2°F of warming because of previous greenhouse gas emissions, and it is difficult to imagine that rapidly growing economies in places like China and India will not significantly contribute to atmospheric CO_2 for the foreseeable future.[15] The Keeling Curve promises to continue its undulating path upward.

Even so, as Stephen Schneider took pains to highlight in the last years of his life, the human and environmental impacts of a changing climate vary significantly depending on the rate and magnitude of change.[16] CO_2 and temperature will likely continue to rise; but as the Supreme Court hinted in *Massachusetts v. EPA*, the nature and extent of that rise is contingent upon how much CO_2 (and its equivalents) the world emits. There are many paths that we can take. Emissions scenarios are bound on the low end only by present conditions and on the high end only by the availability of greenhouse gas sources and the plausibility of worst cases. The potential impacts of scenarios vary accordingly from the low end to the high end. Many of the nonlinear impacts associated with rapid or severe warming (like the collapse of the West Antarctic Ice Sheet) lie in the future, and emissions reductions can decrease their likelihood or potential severity.[17] Rhetorically, for a climate change advocacy group like 350.org, a global CO_2 concentration of 351 ppm in 2050 represents a failure to ensure climatic stability. But 351 ppm in 2050 looks very different from 500 ppm in 2050. With some amount of warming already guaranteed, "success" in preventing global warming altogether remains a chimera, but failure has many flavors, some far more palatable than others.

Recognizing this variability of failure is one of the things that makes climate action plans so compelling. Individual communities are not going to "solve" global warming. Even communities, cities, and states working together to meet Kyoto Protocol standards stand little chance of stabilizing greenhouse gases in the way that national and international targets and timetables are meant to do. But communities across the United States have gone ahead with climate action plans anyhow, because they have engaged in thoughtful and often difficult moral, social, and political conversations that have led them to believe that taking action locally is the right thing to do. They have taken the UNFCCC's rhetoric of equity and "common but differentiated responsibilities" seriously, deciding that they have a moral responsibility to live their lives differently regardless of what other nations or communities do.[18]

It is this moral engagement that has been missing from the top-down, science-first approach to global warming advocacy. Mayors, city council members, and governors have certainly used science to establish their positions on global warming and to promote climate action plans to their constituents. Science and economics will always be important in describing the problem of global warming and evaluating the actions we take to deal with it. As a practical matter, responding to global warming requires these objective terminologies. But the models and curves of climate science have also tended to mask the moral, social, and political questions that global interdependence raises about a comfortable and wealthy fossil fuel-driven lifestyle that ultimately carries social and environmental costs for the rest of the globe. In twenty-six states and dozens of municipalities across the country, citizens have decided to engage with such extrascientific questions as they discuss what they hope their regions will look like in the future.[19] They have, in a sense, come out from behind the curve. These local, value-driven plans do not automatically privilege the global environment over other local, state, and national interests—nor necessarily should they—and the patchwork of climate change policy they create is porous and confusing. But these plans carry the political weight of a morally engaged constituency willing to make hard decisions about global warming in the places where they live. Amid the depressing and desperate realities of global warming history and contemporary climate change policy, it is this growing willingness to engage in difficult moral discussions about global warming at home that gives me hope.

NOTES

INTRODUCTION

1   See Scripps Institution of Oceanography, "The Keeling Curve: A Daily Record of Atmospheric Carbon Dioxide from Scripps Institution of Oceanography at UC San Diego," http://keelingcurve.ucsd.edu.

2   Ibid.

3   John Tyndall, "On the Absorption and Radiation of Heat by Gases and Vapours, and on the Physical Connection of Radiation, Absorption, and Conduction," *Philosophical Magazine* 4:22 (1861): 169–94, 273–85; John Tyndall, "On Radiation through the Earth's Atmosphere," *Philosophical Magazine* 4:125 (1862): 200–206; John Tyndall, "Further Researches on the Absorption and Radiation of Heat by Gaseous Matter" (1862), in *Contributions to Molecular Physics in the Domain of Radiant Heat* (New York: Appleton, 1873), 69–121. For more on Tyndall's work on CO_2, see Joshua P. Howe, "Getting Past the Greenhouse: John Tyndall and the Nineteenth-Century History of Climate Change," in *The Age of Scientific Naturalism: John Tyndall and His Contemporaries*, ed. Bernard Lightman and Michael Reidy (London: Pickering and Chatto, 2014) pg. 33–49; James Rodger Fleming, *Historical Perspectives on Climate Change* (Oxford: Oxford University Press, 1998).

4   Svante Arrhenius, "On the Influence of Carbonic Acid in the Air upon the Temperature of the Ground," *London, Edinburgh, and Dublin Philosophical Magazine and Journal of Science* ser. 5 (Apr. 1896): 237–76.

5   For a book that does do this, and does it very well, see Paul Edwards, *A Vast Machine: Computer Models, Climate Data, and the Politics of Global Warming* (Cambridge, MA: MIT Press, 2010).

6   Scripps Institution of Oceanography, "The Keeling Curve"; NOAA, "Carbon Dioxide, Methane Rise Sharply in 2007," April 23, 2008, www.noaanews .noaa.gov/stories2008/20080423_methane.html.

7   See, for example, Ross Gelbspan's *Boiling Point: How Politicians, Big Oil and Coal, Journalists, and Activists Have Fueled a Climate Crisis—and What We Can Do to Avert Disaster* (New York: Basic Books, 2004) and *The Heat Is On: The Climate Crisis, the Cover-Up, the Prescription* (New York: Basic Books, 1998). See

also Stephen H. Schneider, *Science as a Contact Sport: Inside the Battle to Save Earth's Climate* (Washington, D.C.: National Geographic, 2009).

8 See Naomi Oreskes and Erik M. Conway, "The Denial of Global Warming," in *Merchants of Doubt: How a Handful of Scientists Obscured the Truth on Issues from Tobacco Smoke to Global Warming* (New York: Bloomsbury, 2010), 169–215. See also Naomi Oreskes, Erik M. Conway, and Matthew Shindell, "From Chicken Little to Dr. Pangloss: William Nierenberg, Global Warming, and the Social Deconstruction of Scientific Knowledge," *Historical Studies in the Natural Sciences* 38:1 (Winter 2008): 109–52; Myanna Lahsen, "Experiences of Modernity in the Greenhouse: A Cultural Analysis of a Physicist 'Trio' Supporting the Backlash against Global Warming," *Global Environmental Change* 18 (2008): 204–19; Mark Bowen, *Censoring Science: Inside the Political Attack on Dr. James Hansen and the Truth of Global Warming* (New York: Plume, 2008); Jeremy Leggett, *The Carbon War: Global Warming and the End of the Oil Era* (New York: Routledge, 2001); and James Hoggan, *Climate Cover-Up: The Crusade to Deny Global Warming*, with Richard Littlemore (New York: Greystone Books, 2009). Much more nuanced is Michael Hulme's *Why We Disagree about Climate Change: Understanding Controversy, Inaction, and Opportunity* (Cambridge: Cambridge University Press, 2009).

9 James C. Scott, *Weapons of the Weak: Everyday Forms of Peasant Resistance* (New Haven, CT: Yale University Press, 1985). For a clear articulation of this idea, see Ramachandra Guha, *The Unquiet Woods: Ecological Change and Peasant Resistance in the Himalaya* (Berkeley: University of California Press, 1989), 98.

10 For more on emplotment in climate change history, see my essay "The Stories We Tell," *Historical Studies of the Natural Sciences* 42:3 (2012): 244–54.

11 This 43 percent is called the airborne fraction. See Corinne Le Quéré, Michael R. Raupach, Josep G. Canadell, Gregg Marland, Laurent Bopp, Philippe Ciais, Thomas J. Conway, et al., "Trends in the Sources and Sinks of Carbon Dioxide," *Nature Geoscience* 2 (Nov. 2009): 831–36.

12 Oregon farmers may or may not try to capitalize on longer, warmer, typically drier summers by switching to citrus, but agricultural predictions based on the 2007 assessment models of the Intergovernmental Panel on Climate Change and the earlier Hadley models suggest that climatic conditions could make oranges a possibility. For climate change impacts on the Pacific Northwest and its agriculture, see F. N. Tubiello, C. Rosenzweig, R. A. Goldberg, S. Jagtap, and J. W. Jones, *U.S. National Assessment Technical Report Effects of Climate Change on U.S. Crop Production*, pt. 1, *Wheat, Potato, Corn, and Citrus* (Washington, D.C.: U.S. Department of Agriculture, 2000); P. W. Mote, "Trends in Temperature and Precipitation in the Pacific Northwest," *Northwest Science* 77 (2003): 271–82; and State of Oregon, *The Oregon Climate Change Adaptation Program*, December 2010, www.oregon.gov/ENERGY/GBLWRM/docs/Framework_Final_DLCD.pdf.

13 O. A. Anisimov, D. G. Vaughan, T. V. Callaghan, C. Furgal, H. Marchant,

T. D. Prowse, H. Vilhjálmsson, et al., "Polar Regions (Arctic and Antarctic)," in *Climate Change 2007: Impacts, Adaptation and Vulnerability; Contribution of Working Group II to the Fourth Assessment Report of the Intergovernmental Panel on Climate Change*, ed. M. L. Parry, O. F. Canziani, J. P. Palutikof, P. J. van der Linden, and C. E. Hanson (Cambridge: Cambridge University Press, 2007), 653–85.

14 Edwards, *A Vast Machine*, xv.

15 The literature on objectivity in science is enormous. The list is nearly endless, and ranges across many disciplines, but a few good places to start are Theodore Porter, *Trust in Numbers: The Pursuit of Objectivity in Science and Public Life* (Princeton, NJ: Princeton University Press, 1995); Robert Proctor, *Value-Free Science? Purity and Power in Modern Knowledge* (Cambridge, MA: Harvard University Press, 1991); Loraine Daston and Peter Galison, *Objectivity* (New York: Zone Books, 2007); and Bruno Latour and Steve Woolgar, *Laboratory Life: The Construction of Scientific Facts* (Princeton, NJ: Princeton University Press, 1986).

16 Draft notes of the American Association for the Advancement of Science (AAAS) Advisory Group on Climate meeting, May 26, 1978, AAAS Working Group on Climate meeting (transcript), AAAS Climate Program Records, AAAS Archives, Washington, D.C.

17 For critical histories of global warming, see Spencer Weart, *The Discovery of Global Warming* (Cambridge, MA: Harvard University Press, 2003), and the accompanying website, *The Discovery of Global Warming: A Hypertext History of How Scientists Came to (Partly) Understand What People are Doing to Cause Climate Change*, Feb. 2013, www.aip.org/history/climate/index.htm; Fleming, *Historical Perspectives*; Erik M. Conway, *Atmospheric Science at NASA: A History* (Baltimore: Johns Hopkins University Press, 2008); and Oreskes and Conway, *Merchants of Doubt*. For more popular histories, see William K. Stevens, *The Change in the Weather: People, Weather, and the Science of Climate* (New York: Dell Publishing, 1999); and Gale E. Christianson, *Greenhouse: The 200-Year Story of Global Warming* (New York: Penguin Books, 1999).

18 See Isabel Hyde, "The Tragic Flaw: Is It a Tragic Error?," *Modern Language Review* 58:3 (July 1963): 321–25.

19 Richard Dawkins, "Memes: The New Replicators," in *The Selfish Gene* (Oxford: Oxford University Press, 1976), 189–201.

20 Weart's *Discovery* introduces the Cold War connection well. See also Spencer Weart, "From the Nuclear Frying Pan into the Global Fire," *Bulletin of the Atomic Scientists* 48:5 (June 1992): 18–27; Paul Edwards, "The World in a Machine," in *Systems, Experts, and Computers: The Systems Approach in Management and Engineering, World War II and After*, ed. Agatha Hughes and Thomas Hughes (Cambridge, MA: MIT Press, 2000), 221–53; and Conway, *Atmospheric Science at NASA*.

21 Spencer Weart, "Money for Keeling: Monitoring CO_2 Levels," *Historical Studies in the Physical and Biological Sciences* 37:2 (Mar. 2007): 435–52.

22 Thomas Friedman, *Hot, Flat, and Crowded: Why We Need a Green Revolution—and How It Can Renew America* (New York: Farrar, Straus, and Giroux, 2008).

23 Hulme, *Why We Disagree*, xxxviii–xxxix.

24 Periodically, I do use the term *global warming* to describe climate change in political contexts that predate the term's usage—for example, the title of the first chapter is "The Cold War Roots of Global Warming—but with a few exceptions I have made an effort to relate this history in my characters' terms.

25 Oxford English Dictionary, 2nd ed., s.v. "climate." "Klima" is the title of a recent issue of *Osiris* dedicated to the history of climate change, edited by James Rodger Fleming and Vladimir Jankovic. *Osiris* 26:1 (2011).

26 Lucien Boia, *The Weather in the Imagination*, trans. Roger Leverdier (London: Reaktion Books, 2005). Mark Harrison treats the subject in detail in the context of British colonial expansion in *Climates and Constitutions: Health, Race, Environment and British Imperialism in India, 1600–1850* (Oxford: Oxford University Press, 1999). See also Fleming, *Historical Perspectives*, 4–33.

27 "Global Warming Potentials," in *Contribution of Working Group I to the Fourth Assessment of the Intergovernmental Panel on Climate Change, 2007*, ed. S. Solomon, D. Qin, M. Manning, Z. Chen, M. Marquis, K. B. Averyt, M. Tignor, et al. (Cambridge: Cambridge University Press, 2007), 31–34.

CHAPTER I

1 The appellation *grandfather, godfather,* or *father* of global warming shows up in all sorts of places, from interviews with Revelle to accounts of his life. See Walter Sullivan's obituary, "Roger Revelle, 82, Early Theorist in Global Warming and Geology," *New York Times,* July 17, 1991; Judith Morgan and Neil Morgan, *Roger: A Biography of Roger Revelle* (San Diego: University of California, San Diego—Scripps Institution of Oceanography, 1996); and Fleming, *Historical Perspectives*, 122.

2 Deborah Day, "Roger Randall Dougan Revelle Biography," 2008, Scripps Institution of Oceanography Archives, San Diego, www.scilib.ucsd.edu/sio/biogr/Revelle_Biogr.pdf; Thomas F. Malone, Edward D. Goldberg, and Walter H. Monk, "Roger Randall Dougan Revelle, March 7, 1909–June 15, 1991," in *Biographical Memoirs,* vol. 75 (Washington, D.C.: National Academies Press, 1998), 288–309, http://books.nap.edu/openbook.php?record_id=9649&page=288; Roger R. Revelle, "Preparation for a Scientific Career," oral history by Sarah Sharp, 1984, SIO Reference Series No. 88, Nov. 1988, Regional Oral History Office, Bancroft Library, University of California, Berkeley.

3 Revelle's interest in the carbon cycle dates back to his 1936 University of California, Berkeley, Ph.D. dissertation on carbon deposits in the ocean, "Marine Bottom Samples Collected in the Pacific Ocean by the *Carnegie* on Its Seventh Cruise." See also Day, "Revelle Biography."

4 See Edwards, *Vast Machine*, 207–15; and David M. Hart and David G. Victor,

"Scientific Elites and the Making of U.S. Policy for Climate Change Research, 1957–1974," *Social Studies of Science* 23:4 (Nov. 1993): 648.

5   See Revelle, "Preparation for a Scientific Career," oral history by Sarah Sharp, part 1, Bancroft Library, University of California, Berkeley; Roger Revelle, interview by Earl Droessler, Feb. 1989, American Institute of Physics, College Park, MD; and Gustaf Olof Svante Arrhenius, "Oral History of Gustaf Olof Svante Arrhenius," interview by Laura Harkewicz, Apr. 11, 2006, Scripps Institution of Oceanography Archives, La Jolla, CA, http://libraries .ucsd.edu/locations/sio/scripps-archives/resources/collections/oral.html; and Weart, *Discovery*, 28.

6   The 14 in C^{14} represents the isotope's mass number, comprised of the number of neutrons (8) and the number of protons (6). C^{12} contains 6 neutrons and 6 protons.

7   Hans E. Suess, "Radiocarbon Concentration in Modern Wood," *Science* 122 (Sept. 2, 1955), 415–17; C. D. Keeling, "The Suess Effect: ^{13}Carbon-^{14}Carbon Interrelations," *Environment International* 2 (1979): 229–300. See also "Hans Suess Papers: Background," Register of Hans Suess Papers, 1875–1989, Mandeville Special Collections Library, University of California, San Diego, http://libraries.ucsd.edu/speccoll/testing/html/mss0199a.html; and Heinrich Waenke and James R. Arnold, "Hans E. Suess, December 16, 1909–September 20, 1993," in *Biographical Memoirs*, vol. 87 (Washington, D.C.: National Academies Press, 2005), 354–73, www.nap.edu/openbook.php?record_ id=11522&page=354; Fleming, *Historical Perspectives*, 125; and Weart, *Discovery*, 28–29.

8   Weart, *Discovery*, 28.

9   Roger Revelle and Hans E. Suess, "Carbon Dioxide Exchange between Atmosphere and Ocean and the Question of an Increase in Atmospheric CO_2 during the Past Decades," *Tellus* 9 (1957): 19. This passage, or parts of it, has been quoted as the starting point for scientists' concern over global warming, though as Fleming discusses in detail in *Historical Perspectives*, 107–28, both the subject and the tone of the passage echo earlier statements by other scientists, particularly Guy Stewart Callendar and Gilbert Plass. See Gilbert N. Plass, "The Carbon Dioxide Theory of Climate Change," *Tellus* 7:2 (1956): 140–54. See also Weart, *Discovery*, 30.

10   Gustaf Olof Svante Arrhenius, "Oral History of Gustaf Olof Svante Arrhenius."

11   Ibid.; Revelle and Suess, "Carbon Dioxide Exchange," 19; Roger Revelle, testimony in *National Science Foundation: International Geophysical Year*, hearings of House Committee on Appropriations, 84th Cong., Feb. 6, 1956 (Washington, D.C.: U.S. Government Printing Office, 1956): 473. See also Roger Revelle's testimony in *Report on International Geophysical Year*, hearings of House Committee on Appropriations, 85th Cong., May 1, 1957 (Washington, D.C.: U.S. Government Printing Office, 1957), 113.

12 Hart and Victor, "Scientific Elites," 651.

13 Christianson, *Greenhouse*, 151–57. See also Revelle, interview by Droessler, Feb. 1989, American Institute of Physics; Day, "Revelle Biography"; and Weart, *Discovery*, 36.

14 Weart, *Discovery*, 36.

15 G. M. Woodwell, R. H. Whittaker, W. A. Reiners, G. E. Likens, C. C. Delwiche, and D. B. Botkin, "The Biota and the World Carbon Budget," *Science* 199 (Jan. 13, 1978): 141–46.

16 Daniel Yergin, *The Quest: Energy, Security, and the Remaking of the Modern World* (New York: Penguin, 2011), 439–40.

17 For the founding of IGY, see Water Sullivan, *Assault on the Unknown: The International Geophysical Year* (New York: McGraw-Hill, 1961), 4–35; and Fae L. Korsmo, "The Genesis of the International Geophysical Year," *Physics Today* (July 2007): 38–43.

18 Sullivan, *Assault on the Unknown*, 22; Korsmo, "Genesis," 40.

19 Thomas Malone recalls a fierce battle over changing the name of the *Journal of Meteorology* to the *Journal of Atmospheric Science* in 1960, during his term as president of the American Meteorological Society. The term *atmospheric science*, Malone remembers, came from AAAS president Paul Klopsteg. Thomas F. Malone, interview by Earl Droessler, Feb. 18, 1989, AMS/UCAR Tape Recorded Interview Project, UCAR/NCAR Archives, Boulder, CO.

20 Sullivan, *Assault on the Unknown*, 412, 415.

21 Korsmo, "Genesis," 41.

22 Ibid.

23 On the bomb, see Gregg Herken, *The Brotherhood of the Bomb: The Tangled Lives and Loyalties of Robert Oppenheimer, Ernest Lawrence, and Edward Teller* (New York: Henry Holt, 2002); Kai Bird and Martin J. Sherwin, *American Prometheus: The Triumph and Tragedy of J. Robert Oppenheimer* (New York: Random House, 2005); and Richard Rhodes, *The Making of the Atomic Bomb* (New York: Simon and Schuster, 1986).

24 Report on the International Geophysical Year, hearings of House Committee on Appropriations, 85th Cong., May 5, 1957, 125.

25 Sullivan, *Assault on the Unknown*, 240.

26 Edwards's *Vast Machine* provides an excellent, comprehensive history of climate data and global circulation models.

27 Oxford English Dictionary, 2nd ed., s.v. "weather."

28 Oxford English Dictionary, 2nd ed., s.v. "climate." See also Denis Hartman, *Global Physical Climatology* (San Diego: Academic Press, 1994), 377.

29 See Kristine Harper, *Weather by the Numbers: The Genesis of Modern Meteorology* (Cambridge, MA: MIT Press, 2008). For parallels in oceanography, also at the Bergen school, see Helen Rozwadowski's *The Sea Knows No Boundaries: A Century of Marine Science under ICES* (Seattle: University of Washington

Press, 2002), and *Fathoming the Ocean: The Discovery and Exploration of the Deep Sea* (Cambridge, MA: Harvard University Press, 2005).

30   Harper, *Weather by the Numbers*, 2.

31   See Scott McCartney, *ENIAC: The Triumphs and Tragedies of the World's First Computer* (New York: Walker, 1999); Alice Rowe Burks, *Who Invented the Computer? The Legal Battle That Changed Computing History* (New York: Prometheus Books, 2003); and Eric G. Swedin and David L. Ferro, *Computers: The Life Story of a Technology* (Baltimore: Johns Hopkins University Press, 2007).

32   Harper, *Weather by the Numbers*, 96–97. For more on Zworykin, see Albert Abramson, *Zworykin: Pioneer of Television* (Champaign: University of Illinois Press, 1995).

33   Harper, *Weather by the Numbers*, 141.

34   Edward Lorenz, "Reflections on the Conception, Birth, and Childhood of Numerical Weather Prediction," *Annual Review of Earth and Planetary Sciences* 34 (May 2006): 37–45; Philip Thompson and Edward Lorenz, "Dialogue between Phil Thompson and Ed Lorenz," July 31, 1986, AMS/UCAR Tape Recorded Interview Project, UCAR/NCAR Archives.

35   Joseph Smagorinsky, interview by John Young, May 16, 1986, p. 20, AMS/ UCAR Tape Recorded Interview Project, UCAR/NCAR Archives. For more on how GCMs work and have changed over time, see Edwards, *Vast Machine*, 143–49.

36   G. K. Batchelor, *An Introduction to Fluid Dynamics* (Cambridge: Cambridge University Press, 1967), 147.

37   Hart and Victor, "Scientific Elites," 648.

38   Lester Machta, interview by Julius London, Oct. 31, 1993, p. 5, AMS/UCAR Tape Recorded Interview Project, UCAR/NCAR Archives.

39   Spencer Weart, "Climate Modification Schemes," in *Discovery of Global Warming: A Hypertext History*, www.aip.org/history/climate/RainMake.htm.

40   Weather Modification Research, hearings of House Committee on Interstate and Foreign Commerce, 85th Cong., Mar. 18–19, 1958 (Washington, D.C.: U.S. Government Printing Office, 1958), 9.

41   See Bruce De Silva, "Pell Feels U.S. Wages Weather Warfare," *Providence Journal*, June 26, 1972; and George C. Herring, *America's Longest War: The United States and Vietnam, 1950–1975*, 2nd ed. (New York: McGraw-Hill, 1986), 250.

42   James R. Fleming, "The Climate Engineers: Playing God to Save the Planet," *Wilson Quarterly* (Spring 2007): 55.

43   Ibid., 53.

44   Ibid., 55.

45   Howard T. Orville, "The Impact of Weather Control on the Cold War," in *Weather Modification Research*, hearings of House Committee on Interstate and Foreign Commerce, 85th Cong., Mar. 18–19, 1958, 51–53.

46   See Zuoyue Wang, *In Sputnik's Shadow: The President's Science Advisory*

Committee and Cold War America (New Brunswick, NJ: Rutgers University Press, 2008).

47 N. Rusin and L. Flit, *Man versus Climate*, trans. Dorian Rottenberg (Moscow: Peace Publishers, 1960); *Weather Modification Research*, hearings of House Committee on Interstate and Foreign Commerce, 85th Cong., Mar. 18–19, 1958, 21.

48 Johnson, quoted in *Weather Modification Research*, hearings of House Committee on Interstate and Foreign Commerce, 85th Cong., Mar. 18–19, 1958, 52.

49 National Center for Atmospheric Research and UCAR Office of Programs, "Architecture of the Mesa Laboratory," 2003, http://eo.ucar.edu/what/arch1 .html.

50 Ibid.

51 "Statement by D. W. Bronk," Feb. 5, 1958, in *Preliminary Plans for a National Center for Atmospheric Research: Second Progress Report of the University Committee on Atmospheric Research*, by the National Science Foundation (Washington, DC: National Science Foundation, Feb. 1959) (hereafter Blue Book), C-1, www. ncar.ucar.edu/documents/bluebook1959.pdf. See also Elizabeth Lynn Hallgren, *The University Corporation for Atmospheric Research and the National Center for Atmospheric Research, 1960–1970: An Institutional History* (Boulder, CO: National Center for Atmospheric Research, 1974) (hereafter Green Book), 3.

52 Green Book, 3.

53 Ibid., 4. See also Harper, *Weather by the Numbers*, 7, 76–89.

54 Green Book, 4.

55 Ibid, 3.

56 Blue Book, 6.

57 Green Book, 5.

58 Blue Book, 4; Green Book, 17–22.

59 Alvin Weinberg, "Impact of Large-Scale Science on the United States," *Science* 134:3473 (July 1961): 161–64; Alvin Weinberg, *Reflections on Big Science* (Cambridge, MA: MIT Press, 1967); Peter Galison and Bruce Hevly, eds., *Big Science: The Growth of Large-Scale Research* (Stanford, CA: Stanford University Press, 1992), 357.

60 Blue Book, vii.

61 Bruce Hevly, "Reflections on Big Science and Big History," in Galison and Hevly, *Big Science*, 355–63. For more on "vertical integration," see Robert Kargon, Stuart W. Leslie, and Erica Schoenberger, "Far beyond Big Science: Science Regions and the Organization of Research and Development," in Galison and Hevly, *Big Science*, 335.

62 See Walter Orr Roberts, "Early History of the High Altitude Observatory," July 5, 1966, UCAR Oral History Project, UCAR/NCAR Archives; and Walter Orr Roberts, interview by David H. DeVorkin, July 26–28, 1983, Oral History Program, National Air and Space Museum of the Smithsonian Institution, Washington, D.C.

63 Green Book, 46, citing the NCAR ten-year plan.

64 Ibid., 37.

65 Ibid., 39.

66 For more on atmospheric science at NASA, see Conway, *Atmospheric Science at NASA*. Thanks to Erik Conway for providing me with a copy of his unpublished manuscript for use in the early stages of this work.

67 Blue Book, 39.

68 Ibid., 45.

69 Sea-level rise had been associated with warming at least as early as 1953, when the *New York Times*, citing Gilbert Plass and G. S. Callendar, among others, speculated that continued warming might lead to the inundation of American cities. Leonard Engel, "The Weather Is *Really* Changing," *New York Times Magazine*, July 12, 1953.

70 Report on the International Geophysical Year, hearings of House Committee on Appropriations, 85th Cong., May 5, 1957, 106.

71 Ibid., 108.

72 Bert Bolin and Erik Eriksson, "Changes in the Carbon Dioxide Content of the Atmosphere and Sea Due to Fossil Fuel Combustion," in *The Atmosphere and the Sea in Motion: Scientific Contributions to the Rossby Memorial Volume*, ed. Bert Bolin (New York: Rockefeller Institute Press, 1959), 130–42, available alongside a concise interpretive essay and discussion questions through the National Digital Science Library at http://wiki.nsdl.org/index.php/ PALE:ClassicArticles/GlobalWarming/Article8.

73 As the NDSL essay cited in the prior note briefly suggests, Bolin and Eriksson found inaccuracies in the measured decrease in the ratio of carbon 14 (C^{14}) to carbon 12 (C^{12}) identified by Suess. Again, the reason for the inaccuracies involved the buffering mechanism. Because the oceans take up atmospheric CO_2 slowly and on a limited basis, an increase in atmospheric CO_2 from fossil fuels that have a low C^{14} to C^{12} ratio will be accompanied by a much smaller percentage increase in oceanic CO_2 of the same C^{14} to C^{12} ratio. Because of the existing C^{14}-rich CO_2 in the ocean, the ratio of C^{14} to C^{12} in oceanic CO_2 will be different than that in atmospheric CO_2. "The C14 between the two reservoirs would no longer be in equilibrium," and C^{14} would transfer from ocean to atmosphere, masking the dilution of preexisting, radiocarbon-rich atmospheric CO_2 by new, radiocarbon-poor CO_2 from fossil fuels. Bolin and Eriksson, "Changes in the Carbon Dioxide Content of the Atmosphere and Sea," 131.

74 Rachel Carson, *Silent Spring* (Boston: Houghton Mifflin, 1962). Carson herself followed in a long tradition of American scientists speaking out against environmental degradation, beginning as early as George Perkins Marsh's 1865 *Man and Nature*, with an introduction by David Lowenthal and foreword by William Cronon (Seattle: University of Washington Press, 2003). Among many others, see also William Leroy Thomas, ed., *Man's Role in Changing the*

Face of the Earth (Chicago: University of Chicago Press, 1956); and Paul Sears, Deserts on the March (1937), 4th ed. (Norman: University of Oklahoma Press, 1980).

75 Transcript of *Roger Revelle: Statesman of Science*, a production of KPBS-TV, San Diego, taped Aug. 17, 1992, Box 92, File 39, Scripps Institution of Oceanography Archives, La Jolla, CA. The interview is cited in Day, "Revelle Biography."

76 Day, "Revelle Biography."

77 Sybil P. Parker, "Revelle, Roger," in *McGraw-Hill Modern Scientists and Engineers* (New York: McGraw-Hill, 1980), 20–23, quoted in Day, "Revelle Biography."

78 National Academy of Sciences, Committee on Atmospheric Sciences, *The Atmospheric Sciences, 1961–1971*, vol. 1, *Goals and Plans* (Washington, D.C.: National Academy of Sciences–National Research Council, 1962), 33.

79 Ibid.

80 Conservation Foundation, *Implications of Rising Carbon Dioxide Content of the Atmosphere* (New York: Conservation Foundation, 1963), 14.

81 Air Pollution Control Act of 1955, Pub. L. No. 84-159, ch. 360, 69 Stat. 322 (1955); U.S. Public Health Service, *Clean Air Act, December 17, 1963, as Amended October 20, 1965 and October 15, 1966* (Washington, D.C.: U.S. Government Printing Office, 1967).

82 See J. Fromson, "A History of Federal Air Pollution Control," *Ohio State Law Journal* 30:3 (Summer 1969): 517.

83 See Scripps Institution of Oceanography, "Charles David Keeling Biography," Scripps CO_2 Program, http://scrippsco2.ucsd.edu/sub_program_history /charles_david_keeling_biography.html.

84 Conservation Foundation, *Implications of Rising Carbon Dioxide*, 6–7.

85 Ibid., 4.

86 Ibid., 15.

87 Ibid.

88 U.S. President's Science Advisory Committee (PSAC), Environmental Pollution Panel, *Restoring the Quality of Our Environment: Report of the Pollution Panel, President's Science Advisory Committee* (Washington, D.C.: The White House, 1965).

89 Ibid., 3–9.

90 Ibid., 1.

91 Ibid., 112.

92 The group also included Keeling, Smagorinsky, oceanographer Wallace Broecker, and geochemist Harmon Craig. As Bolin and Eriksson had in 1958, the PSAC carbon dioxide group warned that by the year 2000, the burning of coal, oil, and natural gas could lead to a 25 percent increase in atmospheric CO_2. "This will modify the heat balance of the atmosphere to such an extent that marked changes in climate, not controllable through local or even national efforts, could occur." PSAC, *Restoring*, 112.

93 Ibid., 121.

94 Ibid., 127.

95 G. S. Callendar, "The Artificial Production of Carbon Dioxide and Its Influence on Temperature," *Quarterly Journal of the Royal Meteorological Society* 64 (1938): 223.

96 For more on Callendar and early incarnations of the CO_2 theory of climate, see Fleming, *Historical Perspectives*; and James Rodger Fleming, *The Callendar Effect: The Life and Times of Guy Stewart Callendar (1898–1964), the Scientist Who Established the Carbon Dioxide Theory of Climate Change* (Boston: American Meteorological Society, 2007).

97 Callendar, "Artificial Production," 223.

98 Ibid.

99 PSAC was the second presidential task force to whom Revelle had introduced the issue of CO_2. The first was a subgroup of President Johnson's Domestic Council, which released a report in 1964. Joseph Fisher, Paul Freund, Margaret Mead, and Roger Revelle, "Notes Prepared by Working Group Five, White House Group on Domestic Affairs," Apr. 4, 1964, President's Committee [White House Group on Domestic Affairs], File 42, Box 20, Roger Revelle Collection MC 6, Scripps Institution of Oceanography Archives, La Jolla, CA.

100 Conway, *Atmospheric Science at NASA*, 65. See also Norman Phillips, "Jule Gregory Charney, Jan. 1, 1917–July 16, 1981," in *Biographical Memoirs*, vol. 66 (Washington, D.C.: National Academies Press, 1995), 80–113, www.nap.edu/html/biomems/jcharney.html.

101 Thomas F. Malone, *Reflections on the Human Prospect* (blog), http://humanprospect-post2.blogspot.com.

102 In 1967, the American Association for the Advancement of Science, an organization often on the progressive side of science activism, identified secrecy in science as one of the most important issues facing the American scientific community in the 1960s. Minutes, AAAS Committee on Science in the Promotion of Human Welfare, May 14, 1967, Papers of the Committee on Science in the Promotion of Public Welfare, AAAS Archives, Washington, D.C.

CHAPTER 2

1 "Johnson Gets Report," *New York Times*, May 4, 1965.

2 Joshua Rosenbloom, "The Politics of the American SST Programme: Origin, Opposition, and Termination," *Social Studies of Science* 11:4 (Nov. 1981): 403–23.

3 Wilson was referring specifically to the Anglo-French Concorde, but the sentiment holds for the American SST. *The Supersonic Transport*, hearings of U.S. Congress Joint Economic Committee, 92nd Cong., Dec. 27–28, 1972 (Washington, D.C.: U.S. Government Printing Office, 1972), 89. See also Andrew Wilson, *The Concorde Fiasco* (Middlesex, U.K.: Penguin Books, 1973).

4   For example, see Alexander P. Butterfield to John C. Whitaker, memorandum, Sept. 23, 1969, John. C. Whitaker Papers, White House Central Files, Staff Member Office Files, Nixon Presidential Materials Project, College Park, MD; and Rosenbloom, "Politics of the American SST Programme," 405.

5   B. K. O. Lundberg, "Supersonic Adventure," *Bulletin of the Atomic Scientists* 21 (Feb. 1965): 29–33; *The Supersonic Transport,* hearings of U.S. Congress Joint Economic Committee, 92nd Cong., Dec. 27–28, 1972.

6   The Supersonic Transport, hearings of U.S. Congress Joint Economic Committee, 92nd Cong., Dec. 27–28, 1972, 63.

7   On McNamara and the SST, see Kenneth Owen, *Concorde and the Americans: International Politics of the Supersonic Transport* (Washington, D.C.: Smithsonian Institution Press, 1997). Also see Rosenbloom, "Politics of the American SST Programme," 406.

8   Rosenbloom, "Politics of the American SST Programme," 411.

9   John W. Finney, "Miscalculations by White House and Labor Helped Defeat Supersonic Transport," *New York Times,* Mar. 25, 1971.

10   Mel Horwitch, *Clipped Wings: The American SST Conflict* (Cambridge, MA: MIT Press, 1982), 74.

11   Ibid., 222; Erik M. Conway, *High Speed Dreams: NASA and the Technopolitics of Supersonic Transportation, 1945–1999* (Baltimore: Johns Hopkins University Press, 2005), 126–27; Owen, *Concorde and the Americans,* 114–15.

12   For more on the Oklahoma City tests, see Conway, *High Speed Dreams,* 121–25; and Horwitch, *Clipped Wings,* 75–79.

13   Horwitch, *Clipped Wings,* 222.

14   Udall, quoted in David Hoffman, "Would the SST Peril U.S. Tranquility?" *Washington Post,* Dec. 20, 1967. See also Don Dwiggings, *The SST: Here It Comes Ready or Not* (New York: Doubleday, 1968), 66

15   "Johnson Signs Bill to Curb Jet Plane Noise and Booms," *New York Times,* July 25, 1968; Conway, *High Speed Dreams,* 137.

16   Butterfield to Whitaker, memorandum, Sept. 23, 1969, John C. Whitaker Papers, Nixon Presidential Materials Project.

17   Ibid.

18   Memorandum for the president, Apr. 24, 1969, John C. Whitaker Papers, Nixon Presidential Materials Project.

19   Memorandum for Environmental Quality File, Aug. 9, 1969, John C. Whitaker Papers, Nixon Presidential Materials Project.

20   Economic Analysis and the Efficiency of Government, pt. 4, *Supersonic Transport Development,* hearings of U.S. Congress Joint Economic Committee, Subcommittee on Economy in Government, 91st Cong., May 7, 11, and 12, 1970 (Washington, D.C.: U.S. Government Printing Office,1970), 904–20, also cited in Conway, *High Speed Dreams,* 142.

21   See Earl Lane, "Physicist Richard Garwin: A Life in Labs and the Halls of

Power," American Association for the Advancement of Science news release, Jan. 17, 2006, www.aaas.org/news/releases/2006/0117garwin.shtml.

22   Conway, *High Speed Dreams*, 144.

23   Finney, "Miscalculations by White House."

24   Economic Analysis and the Efficiency of Government, hearings of U.S. Congress Joint Economic Committee, Subcommittee on Economy in Government, 91st Cong., May 7, 1970, 1000.

25   John Firor, interview by Earl Droessler, June 26, 1990, UCAR/NCAR Oral History Project, UCAR/NCAR Archives.

26   See Conway, *Atmospheric Science at NASA*, 94–144; and Weart, *Discovery*, 99–100.

27   Al Morris, "Visit of Dr. Dixie Lee Ray of NSF," memorandum, Jan. 22, 1963, Philip D. Thompson Papers, UCAR/NCAR Archives; Firor, interview by Droessler, June 26, 1990, UCAR/NCAR Archives.

28   William Kellogg to Walt Roberts, "Possibility of a Satellite Experiment to Determine Ozone by Dave," memorandum, Philip D. Thompson Papers, UCAR/NCAR Archives.

29   National Research Council, Committee on Atmospheric Sciences, *Atmospheric Ozone Studies: An Outline for an International Observation Program; A Report by the Panel on Ozone to the Committee on Atmospheric Sciences National Academy of Sciences, National Research Council, Nov., 1965* (Washington, D.C.: National Academy of Sciences–National Research Council, 1966).

30   Firor, interview by Droessler, June 26, 1990, UCAR/NCAR Archives.

31   Conway, *Atmospheric Science at NASA*, 201.

32   See *FAA Certification of the SST Concorde*, hearings of House Committee on Government Operations, 94th Cong., Feb. 24, July 24, Nov. 13–14, Dec. 9, 12, 1975 (Washington, D.C.: U.S. Government Printing Office, 1975), 93.

33   The report was authored by a consortium that actually named itself the Study of Critical Environmental Problems, and its full title was *Man's Impact on the Global Environment: Assessment and Recommendations for Action; Report on the Study of Critical Environmental Problems (SCEP)* (Cambridge, MA: MIT Press, 1970) (hereafter SCEP); Stuart Auerbach, "Scientists Fear Climate Change by SST Pollution," *Los Angeles Times*, Aug. 2, 1970.

34   SCEP, 16.

35   Ibid., 17. See also William Kellogg, interview by Ed Wolff and Nancy Gauss, Feb. 10, 1980, UCAR/NCAR Oral History Project; and George D. Robinson, interview by Earl Droessler, June 27–28, 1994, AMS/UCAR Tape Recorded Interview Project, both in UCAR/NCAR Archives.

36   Auerbach, "Scientists Fear Climate Change by SST Pollution." For the *New York Times* coverage, see Bayard Webster, "Scientists Ask SST Delay Pending Study of Pollution," *New York Times*, Aug. 2, 1970.

37   Civil Supersonic Aircraft Development (SST), hearings of U.S. Congress House Committee on Appropriations, 92nd Cong., Mar. 1–4, 1971

(Washington, D.C.: U.S. Government Printing Office, 1971), 512–41; Christopher Lyndon, "Experts Assure House SST Would Not Be Harmful," *New York Times*, Mar. 4, 1971.

38   Civil Supersonic Aircraft Development, hearings of U.S. Congress House Committee on Appropriations, 92nd Cong., Mar. 1–4, 1971, 524.

39   FAA Certification of the SST Concorde, hearings of U.S. Congress House Committee on Government Operations, 94th Cong., July 24, Nov. 13–14, Dec. 9, 12, 1975, Feb. 24, 1976, 93.

40   Ibid.; Conway, *High Speed Dreams*, 164.

41   See Lydia Dotto and Harold Shriff, *The Ozone War* (Garden City, NY: Doubleday, 1978), 45; Conway, *High Speed Dreams*, 164; and Harold Johnston, "Reduction of Stratospheric Ozone by Nitrogen Oxide Catalysts from Supersonic Transport Exhaust," *Science* 173:3996 (Aug. 1971): 517–22.

42   Walter Sullivan, "Sorry, but There's Still More to Say on SST," *New York Times*, May 30, 1971.

43   FAA Certification of the SST Concorde, hearings of U.S. Congress House Committee on Government Operations, 94th Cong., July 24, Nov. 13–14, Dec. 9, 12, 1975, Feb., 24, 1976. See also Conway, *Atmospheric Science at NASA*, 205.

44   SCEP, 5.

45   See U.N. Environment Programme, "Declaration of the United Nations Conference on the Human Environment," June 16, 1972, www.unep.org/Documents.Multilingual/Default.asp?DocumentID=97&ArticleID=1503.

46   Samuel P. Hays, *Beauty, Health, and Permanence: Environmental Politics in the United States, 1955–1985* (New York: Cambridge University Press, 1987), 8.

47   SCEP, 5.

48   Carroll L. Wilson and William H. Matthews, eds. *Inadvertent Climate Modification: Report of Conference, Study of Man's Impact on Climate (SMIC), Stockholm* (Boston: MIT Press, 1971) (hereafter SMIC), cover and frontispiece.

49   "Charles David Keeling," Wikipedia, http://en.wikipedia.org/wiki/Charles_David_Keeling.

50   For more on Brand, also see Fred Turner, *From Counterculture to Cyberculture: Stewart Brand, the Whole Earth Network, and the Rise of Digital Utopianism* (Chicago: University of Chicago Press, 2008).

51   See also Amory Lovins, *Soft Energy Paths: Toward a Durable Peace* (New York: Harper Collins, 1979).

52   SMIC, 7.

53   Alison Peterson, "John Firor, 80, Early Voice on Environment, Is Dead," *New York Times*, Nov. 12, 2007, www.nytimes.com/2007/11/12/us/12firor.html.

54   Quoted in "Editorials: SST—Some Answers, Some Questions," *Boulder Daily Camera*, Mar. 15, 1971.

55   See Michael Egan, *Barry Commoner and the Science of Survival* (Cambridge, MA: MIT Press, 2007).

56 Daniel Kevles, *The Physicists: The History of a Scientific Community in Modern America* (Cambridge, MA: Harvard University Press, 1995), 402–3.

57 An edited overview of the speeches and panels conducted during the work stoppage can be found in Jonathan Allen, ed., *March 4: Scientists, Students, and Society* (Cambridge, MA: MIT Press, 1970), xxxii.

58 Denis Hayes, interview by the author, June 2, 2009.

59 "Survey Finds Physicists on the Left," *Physics Today* 25 (Oct. 1972): 61, cited in Kevles, *Physicists*, 405.

60 As president of the AAAS in 1968, NCAR founder Walt Roberts oversaw the second of these resolutions. Walter Orr Roberts, FBI File (FOIA), National Center for Atmospheric Research Archives, Boulder, CO; Department of Transportation press release, Dec. 5, 1970, SST Testimony File, William W. Kellogg, Files, UCAR/NCAR Archives. For "AAAS Resolution: Study of Ecological Effects of Herbicides," "AAAS Resolution: Use of Herbicides in Vietnam: Call for Field Study," "AAAS Resolution: Commending U.S. Government Phase-Out of Herbicides in Vietnam," "AAAS Resolution: Assessment of the Ecological Effects of U.S. Activities in Vietnam," and "AAAS Resolution: Appeal for Cessation of Hostilities in Vietnam," see AAAS, "About AAAS: History and Archives," http://archives.aaas.org/docs/resolutions.php.

61 NASA, for example, began a research program at the Lewis Research Center focusing on lower-emission jet engines. Conway, *Atmospheric Science at NASA*, 208.

62 Proctor, *Value-Free Science?*, 10. For more on objectivity and neutrality in general, see Helen E. Longino, *Science as Social Knowledge: Values and Objectivity in Scientific Inquiry* (Princeton, NJ: Princeton University Press, 1989); and Karl Popper, *Objective Knowledge: An Evolutionary Approach* (Oxford: Clarendon Press, 1971). For the concept of scientific neutrality in environmental affairs, see Stephen Bocking, *Nature's Experts: Science, Politics, and the Environment* (New Brunswick, NJ: Rutgers University Press, 2006).

63 Proctor, *Value-Free Science?*, 10.

64 William Kellogg comments on McDonald draft, n.d., SST Testimony File, William W. Kellogg, Files, UCAR/NCAR Archives.

65 Ibid.

66 Ibid.

67 Kellogg did not personally discount McDonald because of his bias; he simply found McDonald's conclusion unconvincing on the particular issue of the SST. "Jim McDonald," he noted, "would not have raised these worries without having some good arguments to back him up." Kellogg comments on McDonald draft, n.d., SST Testimony File, William W. Kellogg, Files, UCAR/NCAR Archives. It is perhaps telling of Kellogg's respect for McDonald as a scientist that during the SST debate, Kellogg did not mention McDonald's persistent and very public interest in the scientific study

of UFOs. For more on the strange life of James McDonald, see Ann Druffel, *Firestorm: Dr. James E. McDonald's Fight for UFO Science* (Columbus, NC: Wild Flower Press, 2003).

68 According to NSF administrators, half of the foundation's overall expenditures on atmospheric sciences went to NCAR. National Science Foundation Advisory Panel for Atmospheric Sciences, "Report on the National Center for Atmospheric Research," Nov. 1970, UCAR/NCAR Archives.

69 Ibid.

70 Johnston to William Kellogg, Oct. 2, 1972, William W. Kellogg, Files, UCAR/NCAR Archives.

71 SMIC, 4.

72 Ibid.

CHAPTER 3

1 *Oxford English Dictionary,* 2nd ed., s.v. "environment" and "weather."

2 Åström, quoted in Maria Ivanova, "Looking Forward by Looking Back: Learning from UNEP's History," in *Global Environmental Governance: Perspectives on the Current Decade,* ed. Lydia Swart and Estelle Perry (New York: Center for UN Reform Education, 2007), 3.

3 Ivanova, "Looking Forward by Looking Back," 3; "U.N. Group Urges Curb on Pollution," *New York Times,* July 31, 1968; "Thant Urges Concerted Action on Pollution Crisis," *New York Times,* June 24, 1969.

4 "The Proceedings in the U.N.," *New York Times,* Apr. 2, 1969.

5 See Adlai E. Stevenson, "Strengthening the International Development Institutions," speech before the U.N. Economic and Social Council, Geneva, Switzerland, July 9, 1965, available at Adlai Today, www.adlaitoday.org/article .php?id=6. See also Barbara Ward, *Spaceship Earth* (New York: Columbia University Press, 1968); and Kenneth Boulding, "The Economics of the Coming Spaceship Earth," in *Environmental Quality in a Growing Economy: Essays from the Sixth RFF Forum,* ed. Henry Jarrett (Baltimore: Johns Hopkins University Press, 1966), 3–14.

6 See Edwards, "World in a Machine," 221–53.

7 International Council of Scientific Unions (ICSU), Scientific Committee on Critical Problems of the Environment (SCOPE), Commission on Monitoring, *Global Environmental Monitoring: A Report Submitted to the United Nations Conference on the Human Environment, Stockholm 1972* (Stockholm: ICSU-SCOPE, 1971); Donella H. Meadows, Dennis L. Meadows, Jørgen Randers, and William W. Behrens III, *The Limits to Growth: A Report for the Club of Rome's Project on the Predicament of Mankind* (New York: Signet, 1972).

8 See Robert Lilienfeld, *The Rise of Systems Theory: An Ideological Analysis* (New York: John Wiley and Sons, 1978).

9 For more on Forrester and the Club of Rome, see Fernando Elichirigoity,

Planet Management: Limits to Growth, Computer Simulation, and the Emergence of Global Spaces (Evanston, IL: Northwestern University Press, 1999), especially chapters 3–5.

10 Meadows et al., *Limits to Growth*, x.

11 Ozbekhan, quoted in Elichirigoity, *Planet Management*, 76.

12 Meadows et al., *Limits to Growth*, xi.

13 See Jay Forrester, *World Dynamics* (Cambridge, MA: Wright-Allen Press, 1971); Edwards, "World in a Machine," 243–45; and Elichirigoity, *Planet Management*, 57.

14 Meadows et al., *Limits to Growth*, xi.

15 The sales figures are from Edwards, "World in a Machine," 244; and John McCormick, *Reclaiming Paradise: The Global Environmental Movement* (Bloomington: Indiana University Press, 1989), 82.

16 For the one reference to CO_2 in *Limits to Growth*, see 72–73.

17 See Edwards, "World in a Machine."

18 For a list of participants, see SCEP, xvii–xxii.

19 Ibid., xi.

20 Ibid., v–x.

21 Ibid., xii.

22 ICSU, SCOPE, Commission on Monitoring, *Global Environmental Monitoring*, 22.

23 SMIC, xv.

24 Ibid.

25 Schneider, *Science as a Contact Sport*, 3.

26 SCEP, 5.

27 Ibid., 245.

28 Eugene Rabinowitch, "International Cooperation of Scientists," *Bulletin of the Atomic Scientists* 2 (Sept. 1946): 1. See also Joseph Manzione, "Amusing and Amazing and Practical and Military: The Legacy of Scientific Internationalism in American Foreign Policy, 1945–1963," *Diplomatic History* 24:1 (Winter 2000): 23. For more on the relationship between American scientists and the state during the Cold War, see Paul Forman, "Behind Quantum Electronics: National Security as a Basis for Physical Research in the United States," *Historical Studies of the Physical and Biological Sciences* 18:1 (1987): 149–229; Daniel J. Kevles, "Cold War and Hot Physics: Science, Security and the American State," *Historical Studies of the Physical and Biological Sciences* 20:2 (1990): 239–64; Aaron L. Friedberg, "Science, the Cold War, and the American State," *Diplomatic History* 20 (Winter 1996): 107–18; and Gregg Herken, "In the Service of the State: Science and the Cold War," *Diplomatic History* 24:1 (Winter 2000): 107–15.

29 Edwards, "Making Global Data," in *A Vast Machine*, 187–227. See also Clark A. Miller, "Scientific Internationalism in American Foreign Policy: The Case of Meteorology, 1947–1958," in *Changing the Atmosphere: Expert Knowledge and*

Environmental Governance, ed. Clark A. Miller and Paul Edwards (Cambridge: MIT Press, 2001), 167–218.

30   SMIC, 27.

31   See also Meadows et al., *Limits to Growth*, 21, 200. Rene Dubos and Barbara Ward, *Only One Earth: The Care and Maintenance of a Small Planet* (New York: W. W. Norton, 1972), was written specifically for the 1972 U.N. Conference on the Human Environment.

32   Gladwin Hill, "U.S. to Suggest Pollution Board Reform," *New York Times*, Dec. 8, 1970.

33   "Planner of Global Talks: Maurice Frederick Strong," *New York Times*, Sept. 23, 1971.

34   Henry Tanner, "Canadian Added to Candidate List for Thant Job," *New York Times*, Nov. 19, 1971; "Planner of Global Talks: Maurice Frederick Strong," *New York Times*, Sept. 23, 1971; Ivanova, "Looking Forward by Looking Back."

35   See Paul Kennedy, *The Parliament of Man: The Past, Present, and Future of the United Nations* (New York: Random House, 2006); and Stanley Meisler, *United Nations: The First Fifty Years* (New York: Atlantic Monthly Press, 1995).

36   Kennedy, *Parliament of Man*, 143–76; Tanner, "Canadian Added to Candidate List for Thant Job."

37   Hill, "Pollution Board Reform."

38   Gladwin Hill, "U.N. Parley Plans a Global Environment Conference," *New York Times*, Sept. 14, 1971.

39   Hays, *Beauty, Health, and Permanence*; Robert Gottlieb, *Forcing the Spring: The Transformation of the American Environmental Movement* (Washington, D.C.: Island Press, 1993); Hal K. Rothman, *The Greening of a Nation? Environmentalism in the United States since 1945* (Fort Worth, TX: Harcourt Brace, 1998); Kirkpatrick Sale, *The Green Revolution: The American Environmental Movement, 1962–1992* (New York: Hill and Wang, 1993); Philip Shabecoff, *A Fierce Green Fire: The American Environmental Movement*, rev. ed. (Washington, D.C.: Island Press, 2003); Roderick Nash, *Wilderness and the American Mind*, 4th ed. (New Haven, CT: Yale University Press, 2001).

40   See Nash, "The International Perspective," in *Wilderness and the American Mind*, 342–79. As Douglas Weiner shows in his study of Russian environmental politics, *A Little Corner of Freedom: Russian Nature Protection from Stalin to Gorbachev* (Berkeley: University of California Press, 1999), environmental activism in Europe was often just as tied to other political movements as American environmentalism was tied to opposition to the war and the concern over free speech and civil liberties. See Christof Mauch, Nathan Stoltzfus, and Douglas R. Weiner, eds., *Shades of Green: Environmental Activism around the Globe* (Lanham, MD: Rowman and Littlefield, 2006); and Ramachandra Guha, *Environmentalism: A Global History* (New York: Longman Publishing Group, 1999).

41   Ivanova, "Looking Forward by Looking Back," 27.

42   "U.N. Group Urges Curb on Pollution," *New York Times*, July 31, 1968.

43 "Thant Urges Concerted Action," *New York Times*, June 24, 1969.

44 Development and Environment: Report and Working Papers of a Panel of Experts Convened by the Secretary-General of the United Nations Conference on the Human Environment (Founex, Switzerland, June 4–12, 1971) (Paris: Mouton, 1972) (hereafter Founex Report), 1, 4.

45 Wade Rowland, *The Plot to Save the World: The Life and Times of the Stockholm Conference on the Human Environment* (Toronto: Clarke, Irwin and Co., 1973): 47, quoted in Ivanova, "Looking Forward by Looking Back," 29.

46 Founex Report, 1.

47 "Kennan Now Backs an Agency in U.N. on World Pollution," *New York Times*, Mar. 3, 1972.

48 Ibid.

49 Sullivan covered the episode in "The Cry of the Vanishing Whale Heeded in Stockholm," *New York Times*, June 9, 1972. See also Walter Sullivan, "Whaling Halt Urged in Stockholm," *New York Times*, June 10, 1972; and "Postscript to Stockholm," *New York Times*, July 7, 1972.

50 Quotations here and through the end of the paragraph are from Sullivan, "Cry of the Vanishing Whale."

51 Memorandum for the president from John C. Whitaker via John D. Ehrlichman, June 23, 1972; "United Nations Conference on the Human Environment, Stockholm, June 5–16, 1972, Scope Paper," both in John C. Whitaker Papers, Nixon Presidential Materials Project.

52 "Scope Paper," John C. Whitaker Papers, Nixon Presidential Materials Project.

53 Walter Sullivan, "U.N. Parley Endorses Air Monitoring Net," *New York Times*, June 8, 1972.

54 UNEP, "Earthwatch," http://earthwatch.unep.net/about/index.php.

55 Ibid.

56 Thomas Malone, interview by the author, June 4, 2009.

57 "Scope Paper," John C. Whitaker Papers, Nixon Presidential Materials Project.

58 Gladwin Hill, "Plan for Coordinating Unit Backed at Ecology Parley," *New York Times*, June 14, 1972; "Scope Paper," John C. Whitaker Papers, Nixon Presidential Materials Project.

59 Hill, "Plan for Coordinating Unit."

60 Christopher DeMuth to John Whitaker, Nov. 3, 1971; "Talking Points for John Ehrlichman," June 13, 1972, both in John C. Whitaker Papers, Nixon Presidential Materials Project. For more on Train and World Heritage Sites, see Nash, *Wilderness and the American Mind*; John McCormick, *Reclaiming Paradise*; J. Brooks Flippin, *Nixon and the Environment* (Albuquerque: University of New Mexico Press, 2000); and J. Brooks Flippin, *Conservative Conservationists: Russell Train and the Emergence of American Environmentalism* (Baton Rouge: Louisiana State University, 2006).

61 Memorandum for the president from Whitaker via Ehrlichman, June 23, 1972, John C. Whitaker Papers, Nixon Presidential Materials Project.

62 Palme, quoted in Gladwin Hill, "Draft Calls for Ecological Responsibility," *New York Times*, June 7, 1972.

63 Tang, quoted in Gladwin Hill, "China Denounces U.S. on Pollution," *New York Times*, June 10, 1972.

64 Gladwin Hill, "Sense of Accomplishment Buoys Delegates Leaving Ecology Talks," *New York Times*, June 18, 1972; "Statement of Roger P. Hansen before the Secretary of State's Advisory Committee on the United Nations Conference on the Human Environment," Mar. 7, 1972, John C. Whitaker Papers, Nixon Presidential Materials Project.

65 The number comes from a memorandum for the president from Russell E. Train, June 19, 1972, John C. Whitaker Papers, Nixon Presidential Materials Project.

66 "Statement of Roger P. Hansen," Mar. 7, 1972, John C. Whitaker Papers, Nixon Presidential Materials Project.

67 "Talking Points for John Ehrlichman," June 13, 1972, John C. Whitaker Papers, Nixon Presidential Materials Project.

68 Hill, "China Denounces U.S."; Hill, "Sense of Accomplishment."

69 Tang, quoted in Hill, "Sense of Accomplishment."

70 Hill, "Draft Calls for Ecological Responsibility."

71 Tang, quoted in Hill, "China Denounces U.S."

72 Hill, "Plan for Coordinating Unit." The "puppet clique" comment was reported in Hill, "China Denounces U.S."

73 "Talking Points for John Ehrlichman," June 13, 1972, John C. Whitaker Papers, Nixon Presidential Materials Project.

74 Anthony Lewis, "One Confused Earth," *New York Times*, June 17, 1972.

75 Hill, "Plan for Coordinating Unit."

CHAPTER 4

1 The quotations (emphasis in the original) are from Michael McCloskey, "Criteria for International Campaigns," Dec. 1, 1982, International Committee, 1972–83, Meetings and Conferences, Operational Files, Sierra Club International Program Records, Box 3, Folder 19, Sierra Club Archives, Bancroft Library Special Collections, University of California, Berkeley.

2 See Hays, *Beauty, Health, and Permanence.*

3 Sid Perkins, "Cooling Climate 'Consensus' of 1970s Never Was," *ScienceNews* 174:9 (Oct. 25, 2008): 5–6; Thomas C. Peterson, William M. Connolley, and John Fleck, "The Myth of the 1970s Global Cooling Scientific Consensus," *Bulletin of the American Meteorological Society* 89:9 (Sept. 2008): 1325–37.

4 Sukiro Manabe and Richard T. Weatherald, "Thermal Equilibrium of the

Atmosphere with a Given Distribution of Relative Humidity," *Journal of the Atmospheric Sciences* 24:3 (1967): 241–59.

5   Reid A. Bryson and Wayne M. Wendland, "Climatic Effects of Atmospheric Pollution," and J. Murray Mitchell Jr., "A Preliminary Evaluation of Atmospheric Pollution as a Cause of the Global Temperature Fluctuation of the Past Century," in *Global Effects of Environmental Pollution*, ed. S. Fred Singer (New York: Springer-Verlag, 1970), 130–38, 139–55; Reid A. Bryson and Thomas J. Murray, *Climates of Hunger: Mankind and the World's Changing Weather* (Madison: University of Wisconsin Press, 1977), 137.

6   Stephen Schneider, interview by Bob Chervin, Jan. 10–13, 2002, AMS/UCAR Tape Recorded Interview Project, UCAR/NCAR Archives; S. I. Rasool and S. H. Schneider, "Atmospheric Carbon Dioxide and Aerosols: Effects of Large Scale Increases on Global Climate," *Science* 173:3992 (July 9, 1971): 138.

7   For more on models, see Dennis L. Hartmann, "Global Climate Models," in *Global Physical Climatology* (San Diego, CA: Academic Press, 1994), 254–85.

8   Bryson and Murray, *Climates of Hunger*, 137; Mitchell, "Preliminary Evaluation of Atmospheric Pollution," 139–55.

9   Stephen Schneider, *The Genesis Strategy: Climate and Global Survival*, with Lynn E. Mesirow (New York: Plenum Press, 1976), 136.

10   Schneider, interview by Chervin, Jan. 10–13, 2002, UCAR/NCAR Archives.

11   See Stephen H. Schneider, "On the Carbon Dioxide-Climate Confusion," *Journal of the Atmospheric Sciences* 32:11 (Nov. 1975): 2060–66.

12   Wallace S. Broecker, "Climatic Change: Are We on the Brink of a Pronounced Global Warming?," *Science* 189:4201 (Aug. 8, 1975): 460.

13   See, for example, "Changing the Climate," *Economist,* June 4, 1977, 88.

14   See "A Study of Climatological Research as It Pertains to Intelligence Problems," Working Paper of the Office of Research and Development, Aug. 1974, in *The Weather Conspiracy: The Coming of the New Ice Age*, report by the Impact Team (New York: Ballantine Books, 1977), 161–96.

15   Louis M. Thompson, "Weather Variability, Climatic Change, and Grain Production," *Science* 188:4188 (May 9, 1975): 535–41.

16   John Maddox, *The Doomsday Syndrome* (New York: McGraw-Hill, 1972).

17   Schneider, *Genesis Strategy,* 136. See also p. 10.

18   Ibid., xiv.

19   Ibid., xvii.

20   Ibid., xiv.

21   Ibid., 246 (emphasis in the original).

22   The book was so successful that Schneider made four *Tonight Show* appearances in the spring and summer of 1977, one with Steve Allen during Carson's recovery from a pinched nerve in May and three with Carson himself in June, August, and September. Schneider also appeared on the final episode of Dan Rather's CBS special "Who's Who," in which he and Rather were filmed

fishing in the Barker Dam Reservoir in Nederland, Colorado (near Boulder). John Carmody, "The TV Column," Washington Post, May 26, 1977; Don Shirley, "The TV Column," Washington Post July 23, 1977; Stephen Schneider, interview by Bob Chervin, Jan. 10–13, 2002.

23 H. E. Landsberg, "Forum," E.O.S.: Transactions of the American Geophysical Union 58:3 (Mar. 1977): 122. See also H. E. Landsberg, "Review: The Genesis Strategy—Climate and Global Survival," E.O.S.: Transactions of the American Geophysical Union 57:9 (Sept. 1976): 634–35.

24 Landsberg, "Review: The Genesis Strategy," 634. Landsberg cites Joseph Smagorinsky, "Global Atmospheric Modeling and the Numerical Simulation of Climate," in Weather and Climate Modification, ed. Wilmot Hess (New York: Wiley, 1974), 633–86.

25 Landsberg, "Forum," 122.

26 F. Kenneth Hare, "Book Review," Bulletin of the American Meteorological Society 58:8 (Aug. 1976): 1016.

27 Ibid.

28 Schneider, interview by Chervin, Jan. 10–13, 2002, UCAR/NCAR Archives.

29 Michael Glantz, in discussion with the author, Nov. 12, 2007; Schneider, interview by Chervin, Jan. 10–13, 2002, UCAR/NCAR Archives.

30 See Nick Cullather, "Miracles of Modernization: The Green Revolution and the Apotheosis of Technology," Diplomatic History 28:2 (Apr. 2004): 227–54. See also Vandana Shiva, The Violence of Green Revolution: Third World Agriculture, Ecology, and Politics (London: Atlantic Highlands, 1991).

31 For a popular account of the controversial U.S. grain sale to the Soviet Union, see James Trager, The Great Grain Robbery (New York: Ballantine Books, 1975). See also Lester Brown, By Bread Alone, with Erik P. Eckholm (New York: Praeger, 1974), 5; and John C. Whitaker, Striking a Balance: Environment and Natural Resources Policy in the Nixon-Ford Years (Washington, D.C.: American Enterprise Institute for Public Policy Research, 1976).

32 Brown, By Bread Alone, 60. Brown is also quoted in Schneider, Genesis Strategy, 7.

33 Ibid., 67, 4.

34 Two internal CIA reports, A Study of Climatological Research as it Pertains to Intelligence Problems, Working Paper of the Office of Research and Development, Aug. 1974, and Potential Implications of Trends in World Population, Food Production, and Climate, Directorate of Intelligence, Office of Political Research, OPR-401, Aug. 1974, were made public as official Library of Congress documents in May of 1976. They were published as appendices in a 1977 alarmist popularization written by a group of "investigative journalists" calling themselves the Impact Team. Weather Conspiracy, 161–96, 197–224. CIA grain export estimates appear on p. 208. See also Deborah Shapley, "The Genesis Strategy (Review)," New York Times, July 18, 1976.

35 The CIA paper was titled "Climatological Research as It Pertains to

Intelligence Problems." See "A Study of Climatological Research," in *Weather Conspiracy*, by Impact Team, 163.

36   Ibid., 166.

37   "Dear Senator," letter, June 13, 1973, U.N. Environment Fund 1973, Sierra Club International Program Records, Box 4, Folder 31; Judith Campbell Bird to Citizen Activists, "United Nations Environmental Programme Threatened by Proposed Cut in U.S. Contribution," memorandum, Apr. 18, 1979, U.N. Environment Program Funding, Sierra Club International Program Records, Box 5, Folder 1; Jacob Scherr (NRDC) to Patricia Scharlin (SC), Apr. 9, 1979, U.N. Environment Program Funding, Sierra Club International Program Records, Box 5, Folder 1, all in Sierra Club Archives.

38   Strong wrote to Scharlin under her maiden name, Rambach. She married during her tenure as the head of the Sierra Club's International Program, and her name appears alternatively as Rambach and Scharlin throughout the Sierra Club International Program records. Maurice F. Strong to Patricia Rambach, Nov. 14, 1975, Operational Records, International Committee, 1972–83, Sierra Club International Program Records, Box 3, Folder 16, Sierra Club Archives.

39   "Sierra Club International Committee Questionnaire—Five Year Plan," n.d., Five Year Plan, 1976–77, Operational Records, Sierra Club International Program Records, Box 3, Folder 10, Sierra Club Archives.

40   Ibid. (emphasis in original).

41   Ibid.

42   Ibid.

43   "Table 1: Question 1A—Rank in Priority of Importance the 6 International Environmental Issues You Wish the Club Should Tackle," in "Sierra Club International Committee Questionnaire—Five Year Plan," n.d., Five Year Plan, 1976–77, Operational Records, Sierra Club International Program Records, Box 3, Folder 10, Sierra Club Archives.

44   "Earthcare Center," Conversations on Earthcare Center Information Center/Network 8/15–16, 1979, Program Goals, Operational Records, Sierra Club International Program Records, Box 3, Folder 14, Sierra Club Archives.

45   Ibid.

46   "Earthcare Center," Sierra Club International Program Records, Box 3, Folder 14, Sierra Club Archives.

47   Ibid.; "Sierra Club Priorities: Board Sets 1980 Conservation Priorities," International Committee, Operational Records, Sierra Club International Program Records, Box 3, Folder 18, Sierra Club Archives.

48   GARP (National Academy of Sciences, U.S. Committee for the Global Atmospheric Research Program), *Understanding Climatic Change: A Program for Action* (Washington, D.C.: National Academy of Sciences; Detroit, MI: Grand River Books, 1975), cited in Weart, *Discovery of Global Warming: A Hypertext History*, www.aip.org/history/climate/index.html.

49 National Research Council, Committee on Climate and Weather Fluctuations and Agricultural Production, Board on Agricultural and Renewable Resources, *Climate and Food: Climatic Fluctuations and U.S. Agricultural Production* (Washington, D.C.: National Academy of Sciences, 1976).

50 National Research Council, Panel on Water and Climate, *Climate, Climatic Change, and Water Supply* (Washington, D.C.: National Academy of Sciences, 1977).

51 National Climate Program Act, hearings of House Committee on Science and Technology, 94th Cong., May 18–20, 25–27, 1976 (Washington, D.C.: U.S. Government Printing Office, 1976).

52 National Climate Program Act, hearings of Senate Committee on Commerce, Science, and Transportation, 95th Cong., June 8–10 and July 5, 6, and 8, 1977 (Washington, D.C.: U.S. Government Printing Office, 1977), 166.

53 Ibid.

54 Text of S. 1652, May 18, 1977, in *National Climate Program Act*, hearings of Senate Committee on Commerce, Science, and Transportation, 95th Cong., June 8–10 and July 5, 6, and 8, 1977, 15.

55 Ibid., 14–33.

56 Ibid., 107.

57 In "The Organization of Scientific Men," *Scientific Monthly* 14:6 (June 1922): 568–78, J. McKeen Cattell characterizes the difference between the AAAS and the NAS as a difference between the scientific House of Commons and the scientific House of Lords. In the mid-1970s, this analogy largely held true (as it does today). When AAAS executive officer William Carey described the AAAS as a "populist association" at a meeting in 1978, however, perhaps he overstated the case. "Meeting on Possible AAAS Role in Climate," Mar. 15, 1978, AAAS Climate Program Records, AAAS Archives.

58 The Committee on Future Directions included, among others, former AAAS president Margaret Mead and planetary astronomer Carl Sagan. "Progress Report on the Committee on Future Directions," Feb. 8, 1978, AAAS Board Materials for the Committee on Climate, AAAS Climate Program Records, AAAS Archives.

59 Revelle served as the organization's president in 1974. "AAAS Committee on Climate," n.d. [1978], AAAS Board Materials, AAAS Archives.

60 Draft notes of AAAS Advisory Group on Climate meeting, May 26, 1978, AAAS Working Group on Climate Meeting (transcript), AAAS Climate Program Records, AAAS Archives. There are actually two draft transcripts of the same meeting in the AAAS Archives, one with the names of speakers and myriad margin notes; and one, very similar, in paragraph form with no names or marginalia. I have used the documents in tandem and almost interchangeably and cite them together as written at the start of this note.

61 Draft notes of AAAS Advisory Group on Climate meeting, May 26, 1978, AAAS Archives.

62 Ibid.

63 Notes of a meeting on possible AAAS role in climate, Mar. 15, 1978, AAAS Board Materials; draft notes of AAAS Advisory Group, May 26, 1978, both in AAAS Archives.

64 Progress report on the Committee on Future Directions, Feb. 8, 1978, AAAS Board Materials, AAAS Archives; draft notes of AAAS Advisory Group, May 26, 1978, AAAS Archives.

65 Draft notes of AAAS Advisory Group on Climate meeting, May 26, 1978, AAAS Archives.

66 Ibid.

67 Ibid. (emphasis in the original).

68 National Academy of Sciences, Climate Research Board, *Carbon Dioxide and Climate: A Scientific Assessment* (Washington, D.C.: National Academy of Sciences, 1979), cited in Oreskes, Conway, and Shindell, "Chicken Little," as "Memo, Climate Research Board, Assembly on Mathematical and Physical Sciences, National Academy of Sciences, 'An Evaluation of the Evidence for CO_2-Induced Climate Change,' NAS AMPS, Film Label: CO_2 and Climate Change: Ad Hoc: General."

69 See Oreskes, Conway, and Shindell, "Chicken Little"; and Oreskes and Conway, "The Denial of Global Warming," in *Merchants of Doubt*, 169–215.

70 See Sam Hays, "The Politics of Science," in *Beauty, Health, and Permanence*, 329–62. See also Bocking, *Natures Experts*.

71 See John C. Barrow, "An Age of Limits: Jimmy Carter and the Quest for a National Energy Policy," in *The Carter Presidency: Policy Choices in the Post–New Deal Era*, ed. Gary M. Fink and Hugh Davis Graham (Lawrence: University Press of Kansas, 1998), 158–223; and James T. Patterson, *Restless Giant: The United States from Watergate to Bush v. Gore* (New York: Oxford University Press, 2005), 125–26.

72 Oreskes, Conway, and Shindell, "Chicken Little," 115.

73 Before the DOE began collaborating with the AAAS on the Annapolis conference, the agency held a conference of its own at Oak Ridge National Laboratory—a highly controversial nuclear research facility. Oreskes and Conway, *Merchants of Doubt*, 171.

74 David Slade to David Burns, Nov. 11, 1978, AAAS Climate Program Records, AAAS Archives.

75 National Climate Program Act, hearings of Senate Committee on Commerce, Science, and Transportation, 95th Cong., June 8–10 and July 5, 6, and 8, 1977, 466.

76 "Draft: Sierra Club International Committee Meeting of June 13–14, New York/Tarrytown (1980)," Meetings and Conferences, International Committee, 1972–83, Sierra Club International Program Records, Box 3, Folder 18, Sierra Club Archives.

77 Michael Glantz, "A Political View of CO_2," discussion paper presented at the

DOE–AAAS Workshop on Environmental and Societal Consequences of Possible CO_2-induced Climate Change, Annapolis, MD, Apr. 1979, AAAS Climate Program Records, AAAS Archives.

78 Samuel P. Hays, *Conservation and the Gospel of Efficiency: The Progressive Conservation Movement, 1890–1920* (Cambridge, MA: Harvard University Press, 1959).

79 Ibid.

80 Michael McCloskey to Patricia Rambach, Aug. 15, 1972, Stockholm Conference on the Human Environment, 1972–73, Subject Files, 1972–85, Sierra Club International Program Records, Box 5, Folder 20, Sierra Club Archives.

81 Hays, *Conservation and the Gospel of Efficiency.*

82 Verner Suomi, introduction to National Academy of Sciences, Climate Research Board, *Carbon Dioxide and Climate: A Scientific Assessment,,* vii, quoted in Oreskes, Conway, and Shindell, "Chicken Little."

83 "Action Flow, U.S. Carbon Dioxide Research and Assessment Program," in David H. Slade, Workshop Correspondences, 1979 AAAS/DOE Workshop File, AAAS Archives.

84 "Costs and Benefits of Carbon Dioxide," *Nature* 279 (May 3, 1979): 1.

85 Slade to Burns, Oct. 2, 1980, AAAS Climate Program Records, AAAS Archives.

CHAPTER 5

1 George Woodwell's work on global forests drove much of the topical overlap between scientists' and environmentalists' interests early in the decade. See Woodwell et al., "Biota and the World Carbon Budget"; and George Woodwell, "The Carbon Dioxide Question," *Scientific American* 238:1 (Jan. 1978): 34–43. See also Weart, *Discovery,* 103–7.

2 Michael Schaller, *Reckoning with Reagan: America and Its President in the 1980s* (Oxford: Oxford University Press, 1980), 50. See also "The Reagan Antienvironmental Revolution" in Hays, *Beauty, Health, and Permanence,* 491–526.

3 Schneider, *Science as a Contact Sport,* 90. See also *Fiscal Year 1983 Department of Energy Budget Review (Fossil Energy and Basic Research),* hearings of House Committee on Science and Technology, 97th Cong., Mar. 18, 23, 24, 25, 30, 1982 (Washington, D.C.: U.S. Government Printing Office, 1982), 1.

4 Schneider, *Science as a Contact Sport,* 90. In a 1982 hearing, an exasperated Congressman James Scheuer lamented, "They have cut us down to $4 million, approximately the cost of two M-1 tanks that don't work very well, approximately one quarter of one day's military budget." *Carbon Dioxide and Climate: The Greenhouse Effect,* hearings of House Committee on Science and Technology, 97th Cong., Mar. 25, 1982 (Washington, D.C.: U.S. Government Printing Office, 1982), 7. The same House committee had held hearings the previous summer, on July 31, 1981, under the same title.

5   Denis Hayes, interview by the author, June 6, 2009; Denis Hayes, *Rays of Hope: The Transition to a Post-petroleum World* (New York: W. W. Norton, 1977).

6   National Renewable Energy Laboratory, *25 Years of Research Excellence, 1977–2002* (Boulder, CO: NREL, 2002).

7   Hayes, interview by the author, June 6, 2009; Solar Energy Research Institute, *A New Prosperity: Building a Sustainable Future* (Andover, MA: Brickhouse Publishing, 1981). See also Eric Hirst, "Review: *A New Prosperity: Building a Sustainable Energy Future* by Solar Energy Research Institute," *Policy Sciences* 14:2 (Apr. 1982): 198–202.

8   Hirst, "Review," 199.

9   Hayes, interview by the author, June 6, 2009.

10 Ibid.

11 Ibid.

12 Like many of the scientists and environmentalists I interviewed, Hayes still gets visibly upset talking about Edwards's appointment. "It quickly became clear that Charles Duncan was actually pretty competent," Hayes remembers, "I mean, Jim Edwards was just fucking unbelievable." Ibid.

13 Ibid.

14 Ibid.; Hirst, "Review," 199.

15 Hayes, interview by the author, June 6, 2009.

16 James Watt to the president, "Organization of Energy Functions," Dec. 10, 1981, Department of Energy, 1981–82, Box 3, Folder 2, George A. Keyworth III Files, White House Staff and Office Collections, Ronald Reagan Archives, Ronald Reagan Presidential Library, Simi Valley, CA.

17 Ibid.; Schneider, *Science as a Contact Sport*, 90.

18 Department of Energy Budget Review (Fossil Energy and Basic Research), hearings of House Committee on Science and Technology, 97th Cong., Mar. 18, 23, 24, 25, 30, 1982, 1.

19 In *Science as a Contact Sport*, Stephen Schneider claims that during Slade's presentation of climate study to the incoming DOE officials, James Edwards leaned over to his deputy director of research, N. Douglas Pewitt, and said, loudly enough to be heard, "We were put in here to get rid of environmental and social science research, so just forget this project" (90). The source of the "ear-witness" account remains a mystery, but Pewitt's actions in public hearings in 1981 and 1982 seem to support the claim; he made every effort to diminish the status of and funding for DOE research into CO_2 and other environmental issues.

20 Schneider, *Science as a Contact Sport*, 87; *Global Climate Change Prevention Act of 1989*, hearings of Senate Committee on Agriculture, Nutrition, and Forestry, 101st Cong., Nov. 6, 1989 (Washington, D.C.: U.S. Government Printing Office, 1989). Special thanks to Heather West for her undergraduate research on this topic in her unpublished paper "Agriculture, Climate Change, and Ronald Reagan" (Stanford University, June 8, 2007).

21　Initially, the president hoped to disband the CEQ, but the move would have eliminated a cabinet post, which was unacceptable politically, and eliminating the CEQ technically would have violated the National Environmental Policy Act. See Malcolm Baldwin, "Memo for the Vice President and James Baker, III," Feb. 21, 1981; F. Khekouri to Frank Hodsell, "Budget: Council on Environmental Quality," Mar. 5, 1981; and Alan Hill to Frank Hodsell, memorandum, Apr. 16, 1981, all in James A. Baker III Files, Council on Environmental Quality, Box 1, Folder 2, Ronald Reagan Archives. See also Hays, *Beauty, Health, and Permanence,* 494; Tom Delaney, "Memorandum for Al Hill," Mar. 12, 1981, James A. Baker III Files, Council on Environmental Quality, Box 1, Folder 2, Ronald Reagan Archives; and Lance Gay, "Environmental Programs Facing Deep Budget Cuts," *Washington Star,* Mar. 11, 1981.

22　Gerald O. Barney, ed., *Global 2000 Report to the President: Entering the 21st Century* (Washington, D.C.: U.S. Government Printing Office, 1980).

23　Hays, *Beauty, Health, and Permanence,* 491.

24　Sale, *Green Revolution,* 51; Hays, *Beauty, Health, and Permanence,* 494; Gottlieb, *Forcing the Spring,* 292–96.

25　Hays, *Beauty, Health, and Permanence,* 494; Sale, *Green Revolution,* 50. See also Jonathan Lash, *A Season of Spoils: The Reagan Administration's Attack on the Environment* (New York: Pantheon Books, 1984).

26　George Will, "The Robespierre of Western Resistance," *Washington Post,* Jan. 15, 1981.

27　"James Watt and The Puritan Ethic," *Washington Post,* May 24, 1981. The *Post* article attributes the quotation to the *Wall Street Journal* (n.d.).

28　Peterson, quoted in Andy Pasztor, "Watt Tells Senate He Would Balance Use, Protection of Resources as Interior Chief," *Wall Street Journal,* Jan. 8, 1981. See also *Briefing by the Secretary of the Interior,* hearings of House Committee on Interior and Insular Affairs, 97th Cong., Feb. 5, 1981 (Washington, D.C.: U.S. Government Printing Office, 1981); and *James G. Watt Nomination,* hearings of Senate Committee on Energy and Natural Resources, 97th Cong., Jan. 7–8, 1981 (Washington, D.C.: U.S. Government Printing Office, 1981).

29　See, for example, *James G. Watt Nomination,* hearings of Senate Committee on Energy and Natural Resources, 97th Cong., Jan. 7–8, 1981; *Environmental Protection Agency Research and Development Posture,* hearings of House Committee on Science and Technology, 97th Cong., Oct. 22, 1981 (Washington, D.C.: U.S. Government Printing Office, 1981); and *Department of the Interior and Related Agencies Appropriations for 1982, Part 5,* hearings of House Committee on Appropriations, 97th Cong., Feb. 25, Mar. 4, 6, 10–12, 16, 20, Apr. 7, 1981 (Washington, D.C.: U.S. Government Printing Office, 1981).

30　James G. Watt Nomination, hearings of Senate Committee on Energy and Natural Resources, 97th Cong., Jan. 7–8, 1981, 74.

31　For example, see ibid., 49–51, 55, 59–61.

32　Secretary Watt's Refusal to Cooperate with Congress, hearings of House

Committee on Appropriations, 97th Cong., Feb. 24, 1982 (Washington, D.C.: U.S. Government Printing Office, 1982).

33 James M. Perry, "Liberals Find Use for Reagan Crowd—As a Rallying Cry," *Wall Street Journal*, Mar. 30, 1981.

34 See *James G. Watt Nomination*, hearings of Senate Committee on Energy and Natural Resources, 97th Cong., Jan. 7–8, 1981, 153–202.

35 James Gustave Speth, telephone interview by the author, June 15, 2009.

36 Effects of Carbon Dioxide Buildup in the Atmosphere, hearings of House Committee on Energy and Natural Resources, 96th Cong., Apr. 3, 1980 (Washington, D.C.: U.S. Government Printing Office, 1980).

37 Ibid., 3.

38 Schneider, *Science as a Contact Sport*, 88; *Carbon Dioxide and Climate: The Greenhouse Effect*, hearings of House Committee on Science and Technology, 97th Cong., July 31, 1981 (Washington, D.C.: U.S. Government Printing Office, 1981).

39 *Carbon Dioxide and Climate*, hearings of House Committee on Science and Technology, 97th Cong., July 31, 1981, 1–82.

40 See *Oversight of the National Climate Program*, hearings of House Committee on Science and Technology, 97th Cong., Mar. 5, 1981, and Mar. 16, 1982 (Washington, D.C.: U.S. Government Printing Office, 1982).

41 Schneider, *Science as a Contact Sport*, 88.

42 *Carbon Dioxide and Climate*, hearings of House Committee on Science and Technology, 97th Cong., Mar. 25, 1982; *Carbon Dioxide and the Greenhouse Effect*, hearings of House Committee on Science and Technology, 98th Cong., Feb. 28, 1984 (Washington, D.C.: U.S. Government Printing Office, 1984).

43 Schneider, *Science as a Contact Sport*, 88–92; *Carbon Dioxide and Climate*, hearings of House Committee on Science and Technology, 97th Cong., July 31, 1981.

44 *Carbon Dioxide and Climate*, hearings of House Committee on Science and Technology, 97th Cong., July 31, 1981; *Carbon Dioxide and Climate*, hearings of the House Committee on Science and Technology, 97th Cong, Mar. 25, 1982; *Carbon Dioxide and the Greenhouse Effect*, hearings of House Committee on Science and Technology, 98th Cong., Feb. 28, 1984.

45 "Synthetic Fuels Called a Peril to the Atmosphere," *New York Times*, July 31, 1981; U.S. Congress, *Carbon Dioxide and Climate*, hearings of House Committee on Science and Technology, 97th Cong., July 31, 1981, 60–67. See also David M. Burns, "Climate and CO_2," *New York Times*, Apr. 17, 1981.

46 "Good Advice on Energy," *Washington Post*, Dec. 28, 1981. Journalists also made a direct connection between the CO_2 problem and Democrats' and environmentalists' continuing criticism of the synthetic fuels program initiated under Carter. See "Synthetic Fuels Called a Peril to the Atmosphere," *New York Times*, Aug. 1, 1981.

47 Anthony Scoville, "Why the U.S. Ignores the Greenhouse Effect," *Not Man*

Apart (newsletter), p. 24, reprinted in *Carbon Dioxide and Climate*, hearings of the House Committee on Science and Technology, 97th Cong, Mar. 25, 1982, 147.

48 *Carbon Dioxide and Climate*, hearings of the House Committee on Science and Technology, 97th Cong, Mar. 25, 1982, 146.

49 "Michael Oppenheimer CV," available at Princeton University Program in Science, Technology, and Environmental Policy, www.princeton.edu/step/people/faculty/michael-oppenheimer. See also Michael Oppenheimer, "To Delay on Global Warming," *New York Times*, Nov. 9, 1983.

50 James Hansen, interview by Spencer Weart at Goddard Institute for Space Studies, New York City, Oct. 23, 2000, Niels Bohr Library and Archives, American Institute of Physics, College Park, MD, www.aip.org/history/ohilist/24309_1.html. For more on NASA's shift to earth studies, and on the subsequent Earth Observation System, see "NASA Atmospheric Research in Transition," "The Quest for a Climate Observing System," and "Mission to Planet Earth: Architectural Warfare" in Conway, *Atmospheric Science at NASA*, 122–53, 198–242, 243–75.

51 Model Zero operated on the order of a thousand square kilometers—huge by the standards of weather modelers, but adequate for studies of climate. Under Hansen, GISS stood at the cutting edge of climate modeling, and during the National Academy of Science's preparation of its seminal 1979 report on CO_2 and climate, committee chair Jule Charney consulted with Hansen repeatedly in order to incorporate the GISS results into the NAS survey of climate research. Hansen, interview by Weart, Oct. 23, 2000, Niels Bohr Library and Archives. See also Conway, *Atmospheric Science at NASA*, 202.

52 Ibid. See also Walter Sullivan, "Study Finds Warming Trend That Could Raise Sea Levels," *New York Times*, Aug., 22, 1981.

53 James Hansen, D. Johnson, A. Lacis, S. Lebedeff, P. Lee, D. Rind, and G. Russell, "Climate Impact of Increasing Carbon Dioxide," *Science* 213:4511 (Aug. 28, 1981): 966, also quoted in Conway, *Atmospheric Science at NASA*, 202.

54 Hansen et al., "Climate Impact," 965.

55 Ibid.

56 Ibid., 966.

57 Hansen, interview by Weart, Oct. 23, 2000, Niels Bohr Library and Archives.

58 Sullivan, "Study Finds Warming Trend."

59 See Hansen, interview by Weart, Oct. 23, 2000, Niels Bohr Library and Archives. See also Michael C. MacCracken, S. B. Idso, J. Hansen, D. Johnson, A. Lacis, S. Lebedeff, P. Lee, et al., "Climatic Effects of Atmospheric Carbon Dioxide [exchange of letters]," *Science* 220:4599 (May 20, 1983): 873–75; S. Ishtiaque Rasool, "On Predicting Calamities," *Climatic Change* 5:2 (June 1983): 201–2–; J. Hansen, D. Johnson, A Lacis, S. Lebedeff, P. Lee, D. Rind, and G. Russel "Reply to Rasool," *Climatic Change* 5:2 (June 1983): 203–4; and Stephen H. Schneider, William W. Kellogg, V. Ramanathan, Conway B. Leovy,

and Sherwood B. Idso, "Carbon Dioxide and Climate [exchange of letters]," *Science* 210:4465 (Oct. 3, 1980): 6–8.

60 Hansen talked at length about his presentation to Koomanoff in his 2000 interview by Spencer Weart. See also Weart, *Discovery,* 143–44; and Conway, *Atmospheric Science at NASA,* 204. For a popular and polemical account of Hansen's relationship with government officials in the 1980s, see Bowen, *Censoring Science.*

61 Conway, *Atmospheric Science at NASA,* 204.

62 National Academy of Sciences, Climate Research Board, *Carbon Dioxide and Climate.*

63 The NAS group actually stated that there was "no reason to believe that these changes will be negligible," an odd double negative and the sort of grammatical equivocation that would come to characterize later reports on global warming. Ibid., vii.

64 Hansen estimated that CO_2 stood at 293 ppm in 1880 and 335 ppm in 1980 and would reach 373 ppm by 2000. Despite a Northern Hemisphere cooling between 1940 and 1970, NASA's dataset showed an overall increase of 0.4°C during the twentieth century, a figure "roughly consistent" with Model II results. Hansen et al., "Climate Impact," 960; Conway, *Atmospheric Science at NASA,* 202.

65 Conway, *Atmospheric Science at NASA,* 205. See also Hansen, interview by Weart, Oct. 23, 2000, Niels Bohr Library and Archives; and Wallace Broecker, interview by Spencer Weart, Lamont-Doherty Earth Observatory, NY, Nov. 14, 1997, Niels Bohr Library and Archives, www.aip.org/history/ohilist/23909_1.html.

66 Jay Keyworth, "Memorandum for Ed Meese," Oct. 23, 1983, Box 3, Environmental Protection Agency Folder, George A. Keyworth III Files, Ronald Reagan Archives. See also Oreskes, Conway, and Shindell, "Chicken Little," 122.

67 National Academy of Sciences, Carbon Dioxide Assessment Committee, *Changing Climate* (Washington, D.C.: National Academy of Sciences, 1983). See also Weart, *Discovery,* 146; and Conway, *Atmospheric Science at NASA,* 205.

68 Oreskes, Conway, and Shindell, "Chicken Little," 124, citing Thomas Schelling et al. to Philip Handler, "Ad Hoc Study Panel on Economic and Social Aspects of Carbon Dioxide Increase," Apr. 18, 1980, p. 11, courtesy of Janice Goldblum, National Academy of Sciences Archives.

69 Phillip Shabecoff, "Haste on Global Warming Trend Is Opposed," *New York Times,* Oct. 21, 1983.

70 Ibid.

71 Keyworth, "Memorandum for Ed Meese," Oct. 23, 1983; Ronald B. Frankum to William S. Heffelfinger, Oct. 6, 1982, Box 3, Department of Energy 1981–82, File 4 of 8, both in George A. Keyworth III Files, Ronald Reagan Archives.

72 Oreskes, Conway, and Shindell, "Chicken Little," 122.

73 Ibid., 136.

74   Stephen Seidel and Dale Keyes, *Can We Delay a Greenhouse Warming?* (Washington, D.C.: Environmental Protection Agency, 1983).

75   Keyworth, "Memorandum for Ed Meese," Oct. 23, 1983; Jay Keyworth, "OSTP Monthly Report for October 1983," memo for Ed Meese, Oct. 28, 1983, Box 6, OSTP Monthly Report, 1982–84, Folder 1 of 4, both in George A. Keyworth III Files, Ronald Reagan Archives. See also Arlen J. Large, "Warming of the Earth Is Met with a Degree of Reassurance," *Wall Street Journal*, Oct. 21, 1983; "How to Live in a Greenhouse," *New York Times*, Oct. 23, 1983; Edgar Tasschdjian, "The Promise of the Greenhouse," *New York Times*, Nov. 2, 1983; Oppenheimer, "To Delay on Global Warming"; Seidel and Keyes, *Can We Delay a Greenhouse Warming?*; and Shabecoff, "Haste on Global Warming," Oct. 21, 1983.

76   Lawrence Badash provides a concise and well-narrated account of the nuclear winter saga from a political perspective in "Nuclear Winter: Scientists in the Political Arena," *Physics in Perspective* 3:1 (Mar. 2001): 76–105. See also Lawrence Badash, *A Nuclear Winter's Tale: Science and Politics in the 1980s* (Cambridge, MA: MIT Press, 2009). For more on nuclear winter in the context of the antinuclear movement, see David S. Meyer, *A Winter of Discontent: The Nuclear Freeze Movement and American Politics* (New York: Praeger, 1990); and Frances Fitzgerald, *Way Out There in the Blue: Reagan, Star Wars, and the End of the Cold War* (New York: Simon and Schuster, 2000), 179–82.

77   Schaller, *Reckoning with Reagan*, 129.

78   Richard Halloran, "Weinberger Expects 6-Month Delay for Renewal of Talks on Arms Curb," *New York Times*, Jan. 7, 1981; Tom Wicker, "Beware of 'Gaposis,'" *New York Times*, Jan. 9, 1981.

79   See Richard Halloran, "Reagan Arms Policy Said to Rely Heavily on Communications," *New York Times*, Oct. 12, 1981; and Schaller, *Reckoning with Reagan*, 127.

80   Richard Halloran, "Pentagon Draws Up First Strategy for Fighting a Long Nuclear War," *New York Times*, May 30, 1982; Richard Halloran, "Weinberger Confirms New Strategy on Atom War," *New York Times*, June 4, 1982; Tom Wicker, "Crossing a Thin Line," *New York Times*, Oct. 21, 1981. See also Caspar Weinberger, *Fighting for Peace: Seven Critical Years in the Pentagon* (New York: Warner Books, 1990).

81   Schaller, *Reckoning with Reagan*, 128; Jonathan Schell, *The Fate of the Earth* (New York: Random House, 1982). See also Meyer, *Winter of Discontent*; Thomas R. Rochon and David S. Meyer, *Coalitions and Political Movements: The Lessons of the Nuclear Freeze* (Boulder, CO: Lynne Rienner, 1997); and Douglas C. Waller, *Congress and the Nuclear Freeze: An Inside Look at the Politics of a Mass Movement* (Amherst: University of Massachusetts Press, 1987).

82   As the International Union for the Conservation of Nature affirmed in 1981, "Peace is a contributory condition to the conservation of nature, just as conservation itself contributes to peace through the proper and ecologically

sound use of natural resources." Memo, Sierra Club Task Force on Environmental Impacts of Military Projects to Chapter Chairs, Conservation Chairs, Group Chairs, RCC Chairs, Council Delegates, BOD, International Committee, n.d. [ca. 1981], Sierra Club International Program Records, Box 3, Folder 19, Sierra Club Archives.

83 Ibid.

84 The *Ambio* special issue was the culmination of a two-year project by the Swedish academy, supported, in part, by the Sierra Club. *Ambio* press conference invitation, Sierra Club International Program Records, Box 9, Folder 30, Sierra Club Archives.

85 Conway, *Atmospheric Science at NASA*, 208; Badash, "Nuclear Winter," 80.

86 Luis W. Alvarez, Walter Alvarez, Frank Asaro, and Helen V. Michel, "Extraterrestrial Causes for the Cretaceous-Tertiary Extinction," *Science* 208:4448 (June 6, 1980): 1095–1108. See also Conway, *Atmospheric Science at NASA*, 207; Luis W. Alvarez, *Alvarez: Adventures of a Physicist* (New York: Basic, 1987); Trevor Palmer, *Perilous Planet Earth: Catastrophes and Catastrophism through the Ages* (New York: Cambridge University Press, 2003); and C. G. Wohl, "Scientist as Detective: Luis Alvarez and the Pyramid Burial Chambers, the JFK Assassination, and the End of the Dinosaurs," *American Journal of Physics* 75:11 (Nov. 2007): 75.

87 Paul J. Crutzen and John W. Birks, "The Atmosphere after a Nuclear War: Twilight at Noon," *Ambio* 11:2/3 (Winter 1982): 114–25.

88 Badash, "Nuclear Winter," 80.

89 Ibid., 79–80; Conway, *Atmospheric Science at NASA*, 209.

90 See Conway, *Atmospheric Science at NASA*, 209.

91 Badash, "Nuclear Winter," 87.

92 Ibid., 86.

93 Both Stephen Schneider and Carl Sagan's biographers, Ray Spangenburg and Kit Moser, argue that Sagan's liberal and politically active new wife, Ann Druyan, pushed him into the political fray more than anyone else. See Schneider, *Science as a Contact Sport*, 97–98; Ray Spangenburg and Kit Moser, *Carl Sagan: A Biography* (Westport, CT: Greenwood Press, 2004). See also Tom Head, ed., *Conversations with Carl Sagan* (Jackson: University Press of Mississippi, 2006); Keay Davidson, *Carl Sagan: A Life* (New York: Wiley, 2000); and Richard P. Turco, "Carl Sagan and Nuclear Winter," in *Carl Sagan's Universe*, ed. Yervant Terzian and Elizabeth Bilson (New York: Cambridge University Press, 1997), 239–46.

94 Conway, *Atmospheric Science at NASA*, 210.

95 Ibid., 210; Badash, "Nuclear Winter," 80.

96 George Woodwell to Stephen Schneider, July 11, 1983, Nuclear Winter Files, Box 27-1 (3 of 4), Folder 2, Stephen Schneider Papers, UCAR/NCAR Archives.

97 See Paul R. Ehrlich, Carl Sagan, Donald Kennedy, and Walter Orr Roberts,

The Cold and the Dark: The World after Nuclear War; Conference on the Long-Term Worldwide Consequences of Nuclear War, Washington, D.C., 1983 (New York: W. W. Norton, 1984); and Paul R. Ehrlich, John Harte, Mark A. Harwell, Peter H. Raven, Carl Sagan, George M. Woodwell, Joseph Berry, et al., "Long-Term Biological Consequences of Nuclear War," *Science* 222:4630 (Dec. 23, 1983): 1293–1300.

98 Donald Kennedy, "Conference on the Long-Term Biological Consequences of Nuclear War: Introduction," draft, Oct. 24, 1983, Nuclear Winter Files, Box 5033 (1 of 4), UCAR/NCAR Archives. In *The Discovery of Global Warming,* Weart actually refers to the event as a press conference rather than a scientific meeting; in truth, it was some combination of the two. Weart, *Discovery,* 143–44; Badash, "Nuclear Winter," 79.

99 Thomas Malone, "The Moscow Link: Opening Remarks," draft, Oct. 26, 1983, Nuclear Winter Files, Box 5033 (1 of 4), UCAR/NCAR Archives.

100 Schneider, *Science as a Contact Sport,* 102.

101 See Ehrlich et al., *Cold and the Dark.*

102 Badash, "Nuclear Winter," 89.

103 For example, see Philip Shabecoff, "Grimmer View is Given of Nuclear War Effects," *New York Times,* Oct. 31, 1983; Terry Atlas, "A Grim View of Nuclear War; Scientists Say It Would Leave Earth Cold, Dark," *Chicago Tribune,* Oct. 31, 1983; "Scientists See Man's Doom in Nuclear Winter," *Los Angeles Times,* Dec. 8, 1983; and "Biologists Paint Icy Picture of How the World Could End," *Washington Post,* Nov. 1, 1983.

104 See Kennedy, "Conference on the Long-Term Biological Consequences of Nuclear War: Introduction," Oct. 24, 1983, UCAR/NCAR Archives. The quote is from Malone, "Moscow Link," Oct. 26, 1983, UCAR/NCAR Archives. See also Carl Sagan, "Nuclear War and Climate Catastrophe: Some Policy Implications," *Foreign Affairs* 62:2 (Winter 1983–84): 257–92; and R. Turco, O. B. Toon, T. Ackerman, J. B. Pollack, and C. Sagan, "Nuclear Winter: Global Consequences of Multiple Nuclear Explosions," *Science* 222:4630 (Dec. 23, 1983): 285.

105 Sagan, "Nuclear War and Climate Catastrophe," 286.

106 Ibid., 283.

107 Ibid., 286 (emphasis in the original).

108 Ibid., 284.

109 Ibid.

110 Conway, *Atmospheric Science at NASA,* 212. See also Oreskes, Conway, and Shindell, "Chicken Little"; and Naomi Oreskes and Erik M. Conway, "Challenging Knowledge: How Climate Change Became a Victim of the Cold War," in *Agnotology: The Making and Unmaking of Ignorance,* ed. Robert N. Proctor and Londa Schiebinger (Stanford, CA: Stanford University Press, 2008), 55–89.

111 See S. Fred Singer, "The Big Chill? Challenging a Nuclear Scenario," *Wall*

Street Journal, Feb. 3, 1984; and Russell Seitz, "Comments and Correspondences," *Foreign Affairs* 63:4 (Spring 1984): 998–99.

112 Singer essentially reprinted his *Wall Street Journal* editorial—or at the very least rehashed its arguments—in *Nature* and *Science* in the ensuing two years; and Sagan, along with a group of colleagues, including Steve Schneider and Starley Thompson, continued to rebut Singer's attacks. See S. Fred Singer, "Is the 'Nuclear Winter' Real?" *Nature* 310 (Aug. 16, 1984): 625; Edward Teller, "Widespread After-Effects of Nuclear War," *Nature* 310 (Aug. 16, 1984): 621–24; and S. Fred Singer, Cresson H. Kearny, R. P. Turco, O. B. Toon, T. P. Ackerman, J. B. Pollack, and C. Sagan, "On a 'Nuclear Winter,'" *Science* 227:4685 (Jan. 25, 1985): 356–444.

113 Schneider, *Science as a Contact Sport*, 100–101; Conway, *Atmospheric Science at NASA*, 211.

114 Schneider, *Science as a Contact Sport*, 97.

115 Weinberger's assistant secretary of defense, Richard Perle, spelled out the unimportance of nuclear winter to administration defense policy in hearings before the Senate Armed Services Committee in October of 1985. Caspar W. Weinberger, "The Potential Effects of Nuclear War on the Climate: A Report to the United States Congress," Mar. 1985, in *Nuclear Winter and Its Implications*, hearings of Senate Armed Services Committee, 99th Cong., Oct. 2–3, 1985 (Washington, D.C.: U.S. Government Printing Office, 1985): 72–90. See p. iii of Weinberger's report for his response; 133–34 of the hearings for Perle's remarks.

116 Starley L. Thompson and Stephen H. Schneider, "Nuclear Winter Reappraised," *Foreign Affairs* 64:5 (Summer 1986): 983. One of Sagan's compelling arguments had to do with the possibility of human extinction, which Schneider and Thompson had been skeptical about since as early as the spring of 1983. Using the "average" time of existence for a "successful" species, Sagan estimated that nuclear weapons, through the generations they prevented, would actually cost somewhere along the lines of five hundred trillion lives over time. Sagan, "Nuclear War and Climatic Catastrophe," 275.

117 Thompson and Schneider, "Nuclear Winter Reappraised," 984.

118 See *Nuclear Winter and Its Implications*, hearings of Senate Armed Services Committee, 99th Cong., Oct. 2–3, 1985.

119 Schneider, *Science as a Contact Sport*, 100.

120 Teller, "Widespread After-Effects," 624.

121 Badash, "Nuclear Winter," 87; Department of the Navy to Estelle H. Rogers (NRDC), Mar. 14, 1984, and J. A. Lyons, "Memorandum to the Chief of Naval Operations: The World after a Nuclear War," Nov. 7, 1983, both in Nuclear Winter Files (Box 3 of 4), Stephen Schneider Papers, UCAR/NCAR Archives.

122 MacCracken's climatic research at Lawrence Livermore, for example, had originally received funding primarily from the DOE, which ran the

nonclassified portion of the lab. The nuclear winter research gave him access to DNA and DOD money and essentially saved his climate program. Schneider's group, too, benefited from DNA money. Stephen Schneider to Michael MacCracken and Cecil Leith, Dec. 23, 1983, DOD/LLNL Proposal Folder; memo to Stephen Schneider from the Office of the Secretary of Defense, May 29, 1984, Defense Science Board Task Force Folder; Bill Hess, memo to Steve Schneider and Bob Dickinson, Dec. 19, 1983, DNA Proposal Folder, all in Nuclear Winter Files, Box 5033 (1 of 4), UCAR/NCAR Archives.

123 Conway, *Atmospheric Science at NASA*, 211.

124 Schneider to MacCracken and Leith, Dec. 23, 1983; memo to Schneider from the Office of the Secretary of Defense, May 29, 1984; Hess, memo to Schneider and Dickinson, Dec. 19, 1983, all in UCAR/NCAR Archives.

CHAPTER 6

1 Ellis B. Cowling, "Acid Precipitation in Historical Perspective," *Environmental Science and Technology* 16:2 (Feb. 1982): 111A.

2 Ibid., 113A. See also Hays, *Beauty, Health, and Permanence*, 352; Robert H. Boyle and R. Alexander Boyle, *Acid Rain* (New York: Schocken Books, 1983); Carolyn Curtis, ed., *Before the Rainbow: What We Know about Acid Rain* (Washington, D.C.: Edison Electric Institute, 1980); and Ernest Y. Yamarella and Randal H. Ihara, eds., *The Acid Rain Debate: Scientific, Economic, and Political Dimensions* (Boulder, CO: Westview Press, 1985).

3 Cowling, "Acid Precipitation," 113A, 115A. See also Swedish Preparatory Committee for the UN Conference on the Human Environment (Bert Bolin, chairman), *Sweden's Case Study for the United Nations Conference on the Human Environment: Air Pollution across National Boundaries* (Stockholm: Norstadt and Sons, 1972).

4 John C. Whitaker to Ken Cole, memorandum, Jan. 6, 1971; "Sulfur Dioxide Emissions Charge," memo for John B. Connally [sic] Jr., secretary of treasury, Nov. 11, 1971; John C. Whitaker, memorandum, Nov. 17, 1970, all in John C. Whitaker Papers, Nixon Presidential Materials Project. The tax was never implemented, in part because the Office of Management and Budget showed that it would work too well, taxing SO_2 emissions out of existence before the program could generate enough revenue to meet Nixon's pro-business political goals.

5 James Gustave Speth, *Red Sky at Morning: America and the Crisis of the Global Environment* (New Haven, CT: Yale University Press, 2004), 52. "Acid rain," Hays wrote in 1987, in a sentence that could have foreshadowed a discussion of global warming (but did not), "was a widely shared problem that galvanized cleaner-air advocates across many interests, providing a potential base for extending the concern for air quality far beyond urban centers." Hays, *Beauty, Health, and Permanence*, 123.

6   Hays, *Beauty, Health, and Permanence,* 160–62.

7   Ibid., 162–63, 352.

8   See Robert Stavins, "What Can We Learn from the Grand Policy Experiment? Lessons from SO_2 Allowance Trading," *Journal of Economic Perspectives* 12:3 (Summer 1998): 69–88.

9   See Hartman, *Global Physical Climatology,* 49, 70–71, 324–27.

10   Paul J. Crutzen, "The Influence of Nitrogen Oxides on the Atmospheric Ozone Content," *Quarterly Journal of the Royal Meteorological Society* 96:408 (Dec. 1970): 320–25, summarized in Harold S. Johnston, "Atmospheric Ozone," *Annual Review of Physical Chemistry* 43 (1992): 4.

11   Halocarbons are carbon-centered molecules bonded to outer atoms of the halogen family, including fluorine, chlorine, bromine, and iodine and sometimes including hydrogen. They affect stratospheric ozone through a process called photodissociation, wherein certain wavelengths of radiation cause halogens to separate from their associated hydrocarbons. The result is loose halogen molecules—most notably chlorine—that essentially steal the third oxygen atom from ozone and then relinquish it to another oxygen atom to create an O_2 molecule, leaving the halogen to catalyze the process all over again. The stability of halogens, meanwhile, allows them to stay in the stratosphere almost indefinitely, causing indefinite destruction of stratospheric ozone.

12   Mario J. Molina and F. Sherwood Rowland, "Stratospheric Sink for Chlorofluoromethanes: Chlorine Atom-Catalysed Destruction of Ozone," *Nature* 249 (June 28, 1974): 810–12. See also See Hartman, *Global Physical Climatology,* 49, 324–27.

13   Speth, *Red Sky at Morning,* 92–93.

14   Erik Conway tells the story of NASA's "discovery" of the ozone hole in *Atmospheric Science at NASA,* 171–73. See also Sharon L. Roan, *Ozone Crisis: The 15 Year Evolution of a Sudden Global Emergency* (New York: Wiley and Sons, 1989), 131; Edward A. Parson, *Protecting the Ozone Layer: Science and Strategy* (New York: Oxford University Press, 2003); 84–85; and J. C. Farman, B. G. Gardiner, and J. D. Shanklin, "Large Losses of Total Ozone in Antarctica Reveal Seasonal ClO_x/NO_x Interaction," *Nature* 315 (May 16, 1985): 207–10.

15   Cowling, "Acid Precipitation," 111A.

16   Ibid., 113A.

17   Speth, *Red Sky at Morning,* 91–92.

18   Ibid., 92–94.

19   Ibid., 94.

20   Ibid., 92–94.

21   The National Toxicology Program was one such effort. See Hays, *Beauty, Health, and Permanence,* 354–55.

22   Quoted in Barney, *Global 2000,* 265. The origin of the quotation about skin cancer is elusive, though it is most likely the 1975 Climate Impacts Assessment

Program report, National Research Council, "Biological and Medical Effects of Nitrogen Oxide Emissions," in *Environmental Impact of Stratospheric Flight: Biological and Climatic Effects of Aircraft Emissions in the Stratosphere*, report by the Climatic Impact Committee for the National Academy of Sciences and the National Academy of Engineering (Washington, D.C.: National Academy of Sciences), 35–50. *Global 2000* instead cites National Research Council, Panel on Nitrates, *Nitrates: An Environmental Assessment* (Washington, D.C.: National Academy of Sciences, 1978), and also takes information on the links between ozone depletion and skin cancer from National Research Council, Committee on Impacts of Stratospheric Change, *Halocarbons: Effects on Stratospheric Ozone* (Washington D.C.: National Academy of Sciences, 1976).

23 Barney, *Global 2000*, 265.

24 Ibid.

25 Andre T. McCloskey, "The Stewardship of Planet Earth: Protecting Our Planet Twenty Years of Defending the Environment," *BIOS* 60:1/2 (Mar.–May 1989): 40; U.S. Environmental Protection Agency, "Government Ban on Fluorocarbon Gases in Aerosol Products Begins October 15," EPA press release, Oct. 15, 1978, www.epa.gov/history/topics/ozone/01.htm.

26 Peter Usher, interview by Shardul Agrawala, Bonn, Germany, Mar. 4, 1997, quoted in Shardul Agrawala, "Context and Early Origins of the Intergovernmental Panel on Climate Change," *Climatic Change* 39:4 (Aug. 1998): 614.

27 See National Academy of Sciences, Geophysics Research Board, *Energy and Climate: Studies in Geophysics* (Washington, D.C.: National Academy of Sciences, 1977); National Academy of Sciences, Climate Research Board, *Carbon Dioxide and Climate*; National Research Council, CO_2/Climate Review Panel, *Carbon Dioxide and Climate: A Second Assessment* (Washington, D.C.: National Academy of Sciences, 1982); and National Academy of Sciences, Carbon Dioxide Assessment Committee, *Changing Climate* (Washington, D.C.: National Academy of Sciences, 1983). See also National Defense University, *Climate Change to the Year 2000: A Survey of Expert Opinion* (Washington, D.C.: National Defense University, 1978).

28 "Statement" in World Climate Program, *Report of the International Conference on the Assessment of the Role of Carbon Dioxide and of Other Greenhouse Gases in Climate Variations and Associated Impacts, Villach, Austria, 9–15 October, 1985,* WMO No. 661 (Geneva: World Meteorological Organization, 1986), available as *SCOPE 29* from the ICSU, www.icsu-scope.org/downloadpubs/scope29, quoted in Jill Jaeger, *Developing Policies for Responding to Climate Change: A Summary of the Discussions and Recommendations of the Workshops Held in Villach (28 Sept.–2 Oct. 1987) and Bellagio (9–13 Nov. 1987), under the Auspices of the Beijer Institute, Stockholm* (Geneva: World Meteorological Organization; Nairobi: United Nations Environment Programme, 1988), 1–2. See also Alan Hecht and Dennis Tirpak, "Framework Agreement on Climate Change: A Scientific and Political History," *Climatic Change* 29:4 (Apr. 1995): 371–402;

Agrawala, "Context and Early Origins," 608; Michael Oppenheimer, "Developing Policies for Responding to Climate Change," *Climatic Change* 15:1 (1989): 1–4; and Schneider, *Science as a Contact Sport*, 124–26.

29   Quoted in Jaeger, *Developing Policies for Responding to Climate Change*, 3.

30   Ibid., also quoted in Agrawala, "Context and Early Origins," 609.

31   Jaeger, *Developing Policies*, i.

32   Ibid.

33   Ibid., v, 22, 27.

34   Ibid., v. Recommendations for international regulatory mechanisms recur throughout the report.

35   Agrawala, "Context and Early Origins," 610. See also World Meteorological Organization, *Proceedings of the World Conference on the Changing Atmosphere: Implications for Global Security, Toronto, Canada, June 27–30*, WMO No. 710 (Geneva: World Meteorological Organization/United Nations Environment Programme, 1988).

36   Agrawala, "Contexts and Early Origins," 610.

37   Jaeger, *Developing Policies*, 2.

38   Agrawala, "Context and Early Origins," 612.

39   Ibid., 612; and Tirpak, "Framework Agreement," 380.

40   Robert Fleagle, "The U.S. Government Response to Climate Change: Analysis and Appraisal," *Climatic Change* 20:1 (1992): 72, quoted in in Weart, *Discovery*, 143.

41   See Hecht and Tirpak, "Framework Agreement," 381.

42   Ibid., 385.

43   Agrawala, "Context and Early Origins," 614.

44   Schneider, *Science as a Contact Sport*, 125.

45   John Rich to Senator Biden, "The Biden Initiative on Global Warming," Sept. 23, 1986, Working Files of Peter W. Galbraith, Committee on Foreign Relations, 96–102nd Cong., RG 46 Record of the U.S. Senate, Box 34, National Archives and Records Administration, Washington, D.C.

46   Rafe Pomerance, "The Dangers from Climate Warming: A Public Awakening," in *The Challenge of Global Warming*, ed. Dean Edwin Abrahamson (Washington, D.C.: Island Press, 1989), 259–69. Pomerance specifically references the Global Climate Protection Act on pp. 264–65. See also Spencer Weart, "Government: The View from Washington, D.C.," in *Discovery of Global Warming: A Hypertext History*, www.aip.org/history/climate/Govt.htm.

47   See, for example, *Global Warming*, hearings of Senate Committee on Environment and Public Works, 99th Cong., Dec. 10, 1985 (Washington, D.C.: U.S. Government Printing Office, 1986); *Ozone Depletion, the Greenhouse Effect, and Climate Change*, hearings of Senate Committee on Environment and Public Works, 99th Cong., June 10–11, 1986 (Washington, D.C.: U.S. Government Printing Office, 1986); and *Implications of Global Warming for Natural Resources*, hearings of House Committee on Interior and Insular Affairs, 100th Cong.,

Sept. 27 in Washington, D.C. and Oct. 17 in San Francisco, 1988 (Washington, D.C.: U.S. Government Printing Office, 1989).

48   Stevens, *Change in the Weather*, 129.

49   For the ecological impacts of the Yellowstone fires, see Linda L. Wallace, ed., *After the Fires: The Ecology of Change in Yellowstone National Park* (New Haven, CT: Yale University Press, 2004). See also Stephen Pyne, "Burning Questions and False Alarms about Wildfires at Yellowstone," *Forum for Applied Research and Public Policy* 4:2 (Summer 1989): 31–39; and Norman L. Christiansen, James K. Agee, Peter F. Brussard, Jay Hughes, Dennis H. Knight, G. Wayne Minshall, James M. Peek, et al., "Interpreting the Yellowstone Fires of 1988," *Bioscience* 39:10 (Nov. 1989): 678–85.

50   See H. Brammer, "Floods in Bangladesh: Geographical Background to the 1987 and 1988 Floods," *Geographical Journal* 156:1 (Mar. 1990): 12–22; and Sale, *Green Revolution*, 71–72.

51   Ozone Depletion, the Greenhouse Effect, and Climate Change, hearings of Senate Committee on Environment and Public Works, 99th Cong., June 10–11, 1986, 36.

52   "The Greenhouse Effect? Real Enough," *New York Times*, June 23, 1988.

53   Stevens, *Change in the Weather*, 129; Conway, *Atmospheric Science at NASA*, 234.

54   James Hansen, testimony in *Greenhouse Effect and Global Climate Change, Part 2*, hearings of the Senate Committee on Energy and Natural Resources, 100th Cong, June 23, 1988 (Washington, D.C.: U.S. Government Printing Office, 1988), 39–41 (quotation p. 39). At the time, journalists quoted both Hansen's testimony and his comments from posthearing interviews, as in two of Philip Shabecoff's articles, "Sharp Cut in Fossil Fuels Is Urged to Battle Shift in Climate," *New York Times*, June 24, 1988, and "The Heat Is On: Calculating the Consequences of a Warmer Planet Earth," *New York Times*, June 26, 1988.

55   Shabecoff, "Sharp Cut in Fossil Fuels."

56   Ibid. For more on Hansen's famous "stop waffling" moment, see Weart, *Discovery*, 155–57; Conway, *Atmospheric Science at NASA*, 233–36; Fleming, *Historical Perspectives on Climate Change*, 134–35; James Hansen, interview by Spencer Weart, at NASA Center in New York, Nov. 27, 2000, Niels Bohr Library and Archives, www.aip.org/history/ohilist/24309_2.html.

57   Hecht and Tirpak, "Framework Agreement," 384.

58   "Planet of the Year, Endangered Earth," cover, *Time*, Jan. 2, 1989.

59   Sale, *Green Revolution*, 72.

60   "George Bush: From the Text of a Speech Delivered Aug. 31 in Erie Metropark, Mich," *New York Times*, Sept. 24, 1988. Bush made similar comments on a number of occasions. See also "The White House Effect (Global Warming)," *Economist*, Feb. 3, 1990; Hecht and Tirpak, "Framework Agreement," 383; and Conway, *Atmospheric Science at NASA*.

61   See Oreskes and Conway, *Merchants of Doubt*, 186–87. See also Oreskes,

Conway, and Shindell, "Chicken Little"; and Conway, *Atmospheric Science at NASA*, 238.

62 See William K. Reilly, "Breakdown on the Road from Rio: Reform, Reaction and Distraction Compete in the Cause of the International Environment, 1993–1994" Arthur and Frank Payne Lecture, Stanford University, Stanford, CA, Mar. 2, 1994, cited in Hecht and Tirpak, "Framework Agreement," 402.

63 Tim E. Walker, "Promoting Air Quality, Resisting Climate Change: The George H. W. Bush Administration and the Economic Precautionary Principle," paper presented to the Annual Conference of the American Society of Environmental Historians, Boise, ID, Mar. 15, 2008.

64 Ibid.

65 Agrawala, "Context and Early Origins," 617.

66 Hecht and Tirpak, "Framework Agreement," 385.

67 Leggett, *Carbon War*, 3.

68 Ibid., 3.

69 Leggett and his colleagues in the environmental community began to pejoratively refer to these scientists and the conservative think tanks—like the Global Climate Coalition, the Marshall Institute, and the Global Climate Council—as the Carbon Club. Ibid., 10–11.

70 Bocking, *Nature's Experts*, 114. For more on "boundary objects" and modeling, see Jeroen van der Sluijs, Josee van Eijndhoven, Simon Shackley, and Brian Wynne, "Anchoring Devices in Science for Policy: The Case of Consensus around Climate Sensitivity," *Social Studies of Science* 28:2 (Apr. 1998): 291–323. See also Simon Shackley and Brian Wynne, "Representing Uncertainty in Global Climate Change Science and Policy," *Science, Technology, and Human Values* 21:3 (Summer 1996): 275–302; and David H. Guston, "Boundary Organizations in Environmental Policy and Science: An Introduction," *Science, Technology, and Human Values* 26:4 (Autumn 2001): 399–408.

71 Leggett, *Carbon War*, 17.

72 Ibid.

73 Hecht and Tirpak, "Framework Agreement," 386–87.

74 Ibid. For an argument for "climate justice" in the developing world, see Anil Agrawal and Sunita Narain, *Global Warming in an Unequal World: A Case of Environmental Colonialism* (New Delhi: Centre for Science and Environment, 1990).

75 For a detailed analysis of the negotiation process, see D. Bodansky, "The U.N. Framework Convention on Climate Change: A Commentary," *Yale Journal of International Law* 18:2 (1993): 451–558. See also Leggett, *Carbon War*, 25–28.

76 See Bocking, *Nature's Experts*, 124. See also Speth, *Red Sky at Morning*, 170.

77 Bocking, *Nature's Experts*, 174. See also Sheila Jasanoff, "Contested Boundaries in Policy-Relevant Science," *Social Studies of Science* 17:2 (May 1987): 195–230; and Sheila Jasanoff, *The Fifth Branch: Science Advisors as Policymakers* (Cambridge, MA: Harvard University Press, 1990).

78 See Oreskes and Conway, *Merchants of Doubt*, 205–9.

79 As Oreskes and Conway write in *Merchants of Doubt*, 206, Bert Bolin actually made a certain conservatism of language a matter of IPCC policy in order to avoid alarmist language that might lead people to discount the body's results. For more on the idea of "conservatism" in scientific consensus, see Robert Proctor, *Cancer Wars: How Politics Shaped What We Know and Don't Know about Cancer* (New York: Basic Books, 1995), 261–65.

80 John Lanchbery and David Victor, "The Role of Science in the Global Climate Negotiations," in *Green Globe Yearbook of International Co-operation and Development*, ed. Helge Ole Bergesen and Georg Parmann (Oxford: Oxford University Press, 1995), 29–40; van der Sluijs et al., "Anchoring Devices," 314.

81 C. Gordon Bray and Richard Stewart, "A Comprehensive Approach to Addressing Potential Global Climate Change," paper presented at an informal U.S. Department of State, Feb. 3, 1990. See also B. A. Ackerman and R. B. Stewart, "Reforming Environmental Law: The Democratic Case for Market Incentives," *Columbia Journal of Environmental Law* 13 (1988): 153–69. Both cited in Hecht and Tirpak, "Framework Agreement," 386, 401.

CHAPTER 7

1 Leggett, *Carbon War*, 1.

2 Ibid., 1–2.

3 See Stephen Kotkin, *Armageddon Averted: The Soviet Collapse, 1970–2000* (Oxford: Oxford University Press, 2008); and Charles S. Maier, *Dissolution: The Crisis of Communism and the End of East Germany* (Princeton, NJ: Princeton University Press, 1997).

4 See U.S. Congress, 105 S. Res. 98, July 29, 1997 (Washington, D.C.: U.S. Government Printing Office, 1997) (also known as Byrd-Hagel). See also *Conditions Regarding U.N. Framework Convention on Climate Change*, hearings of Senate Committee on Foreign Relations, 105th Cong., June 19 and 25, 1997 (Washington, D.C.: U.S. Government Printing Office, 1997).

5 World Commission on Environment and Development, *Our Common Future* (Oxford: Oxford University Press, 1987) (hereafter Brundtland Report), 54.

6 Emma Rothschild, "Forum: The Idea of Sustainability: Introduction," and "Maintaining (Environmental) Capital Intact," *Modern Intellectual History* 8:1 (Apr. 2011): 147–51, 193–212. See also Robert Solow, "On the Intergenerational Allocation of Natural Resources," *Scandinavian Journal of Economics* 88 (Mar. 1986): 141–49; and Ronald Coase, "The Problem of Social Cost," *Journal of Law and Economics* 3 (Oct. 1960): 1–44.

7 Paul Warde, "The Invention of Sustainability," *Modern Intellectual History* 8:1 (Apr. 2011): 153–70.

8 Odd Arne Westad, *The Global Cold War: Third World Interventions and the Making of Our Time* (New York: Cambridge University Press, 2007), 2, 4, 396.

9   See William Appleman Williams, *The Tragedy of American Diplomacy*, 50th anniversary ed. (New York: W. W. Norton, 2009), especially 18–57.

10 See Walter LaFeber, *America, Russia, and the Cold War, 1945–1996*, 8th ed. (New York: McGraw-Hill, 1997), 171.

11 Brundtland Report, 54.

12 Flora Lewis, "Gorbachev Turns Green: Soviets Seize the Moment on Global Environment," *New York Times*, Aug. 14, 1991.

13 Ibid.

14 Ibid. See also testimony of Andrew Rice in *U.S. Policy toward the 1992 U.N. Conference on Environment and Development*, hearings of House Committee on Foreign Affairs, 102nd Cong., Apr. 17, July 24, and Oct. 3, 1991 (Washington, D.C.: U.S. Government Printing Office, 1991), 136.

15 *U.S. Policy toward the U.N. Conference on Environment and Development*, hearings of House Committee on Foreign Affairs and Committee on Merchant Marines and Fisheries, 102nd Cong., Feb. 26–27, and June 21 and 28, 1992 (Washington, D.C.: U.S. Government Printing Office, 1992), 47.

16 *U.S. Policy toward the 1992 U.N. Conference on Environment and Development*, hearings of House Committee on Foreign Affairs, Apr. 17, July 24, and Oct. 3, 1991.

17 Ibid., 117.

18 *U.N. Conference on Environment and Development*, hearings of House Committee on Science, Space, and Technology, 102nd Cong., May 7, 1991 (Washington, D.C.: U.S. Government Printing Office, 1991), 34–35.

19 *U.S. Policy toward the 1992 U.N. Conference on Environment and Development*, hearings of House Committee on Foreign Affairs, Apr. 17, July 24, and Oct. 1991, 3.

20 Ibid., 135. See also Shannon Halpern, *The United Nations Conference on Environment and Development: Process and Documentation* (Providence, RI: Academic Council for the United Nations System, 1992).

21 *U.S. Policy toward the U.N. Conference on Environment and Development*, hearings of House Committee on Foreign Affairs and Committee on Merchant Marines and Fisheries, 102nd Cong., Feb. 26–27, June 21 and 28, 1992, 15.

22 Ibid., 12.

23 Ibid.

24 Ibid.

25 For a detailed discussion of what Vandana Shiva calls "environmental imperialism," see Shiva, "The Greening of the Global Reach," in *Global Ecology: A New Arena for Political Conflict*, ed. Wolfgang Sachs (London: Zed Books, 1993), 249–56.

26 For a more nuanced analysis of the Carter administration's approach to foreign policy, see Gaddis Smith, *Morality, Reason, and Power: American Diplomacy in the Carter Years* (New York: Hill and Wang, 1986).

27 *U.S. Policy toward the 1992 U.N. Conference on Environment and Development*,

hearings of House Committee on Foreign Affairs and Committee on Merchant Marines and Fisheries, 102nd Cong., Feb. 26–27, June 21 and 28, 1992, 12.

28 Ibid., 6.

29 Ibid.

30 Ibid., 12; Howard LaFranchi, "Europe Misses Its Chance for Leading Role at Rio," *Christian Science Monitor*, June 1, 1992.

31 Yergin, *Quest*, 486.

32 Steven Greenhouse, "A Closer Look: Ecology, the Economy and Bush," *New York Times*, June 14, 1992, quoted in Yergin, *Quest*, 486.

33 Michael Weisskopf, "'Outsider' EPA Chief Being Tested," *Washington Post*, June 8, 1992. See also *U.N. Framework Convention on Climate Change*, hearings of Senate Committee on Foreign Relations, 102nd Cong., Sept. 18, 1992 (Washington, D.C.: U.S. Government Printing Office, 1992), 26.

34 U.N. Framework Convention on Climate Change (UNFCCC), "Text of the Convention," 1992, United Nations, http://unfccc.int/key_documents/the_convention/items/2853.php.

35 *U.N. Framework Convention on Climate Change*, hearings of Senate Committee on Foreign Relations, 102nd Cong., Sept. 18, 1992, 39, 50.

36 Ibid., 39.

37 For examples, see Murray Feshbach, *Ecological Disaster: Cleaning Up the Hidden Legacy of the Soviet Regime* (New York: Twentieth Century Fund Press, 1995).

38 UNFCCC, "Text of the Convention," 4.

39 UNFCCC, "Text of the Convention." Annex I and II countries are on pp. 24 and 25.

40 Howard LaFranchi, "Europe Misses Its Chance for Leading Role at Rio," *Christian Science Monitor*, June 1, 1992.

41 C. Boyden Gray and David B. Rivkin Jr., "A 'No Regrets' Environmental Policy," *Foreign Policy* 83 (Summer 1991): 50.

42 Ibid., 49.

43 *U.N. Framework Convention on Climate Change*, hearings of Senate Committee on Foreign Relations, 102nd Cong., Sept. 18, 1992, 32.

44 Walker, "Promoting Air Quality, Resisting Climate Change."

45 LaFranchi, "Europe Misses Its Chance."

46 Ibid.

47 *U.N. Framework Convention on Climate Change*, hearings of Senate Committee on Foreign Relations, 102nd Cong, Sept. 18, 1992, 50.

48 *U.S. Policy toward the U.N. Conference on Environment and Development*, hearings of House Committee on Foreign Affairs and Committee on Merchant Marines and Fisheries, 102nd Cong., Feb. 26–27, and June 21 and 28, 1992, 13.

49 Reilly estimated that the Clean Air Act, the National Energy Strategy, the America the Beautiful Initiative, and other conservation actions would limit

the increase in U.S. carbon dioxide–equivalent emissions to between 1.4 and 6 percent. *U.N. Framework Convention on Climate Change*, hearings of Senate Committee on Foreign Relations, 102nd Cong., Sept. 18, 1992, 38.

50 UNFCCC, "Text of the Convention," 4 (Article III, Principle 3).

51 These thirty-seven signatories are the Annex I countries under the UNFCCC; their targets, confusingly, are listed under Annex B of the Kyoto Protocol, and they are often called the Annex B countries.

52 *Global Climate Change: The Adequacy of the National Plan*, hearings of House Committee on Foreign Affairs, 103rd Cong., Mar. 1, 1993 (Washington, D.C.: U.S. Government Printing Office, 1993), 7.

53 See Al Gore's testimony and subsequent exchange with Senator McConnell in *U.S. Policy toward the 1992 U.N. Conference on Environment and Development*, hearings of House Committee on Foreign Affairs, 102nd Cong., Apr. 17, July 24, and Oct. 1991, 2–35.

54 *U.N. Framework Convention on Climate Change*, hearings of Senate Committee on Foreign Relations, 102nd Cong., Sept. 18, 1992, 32. For contemporary popularizations of this now common argument, see Lester Brown, Christopher Flavin, and Sandra Postel, *Saving the Planet: How to Shape an Environmentally Sustainable Global Economy* (New York: W. W. Norton, 1991); and Albert Gore, *Earth in the Balance: Ecology and the Human Spirit* (New York: Rodale, 1992).

55 See, for example, Patrick J. Michaels, *Sound and Fury: The Science and Politics of Global Warming* (Washington, D.C.: Cato Institute, 1992).

56 *U.N. Framework Convention on Climate Change*, hearings of Senate Committee on Foreign Relations, 102nd Cong., Sept. 18, 1992, 19.

57 See James M. Scott, ed., *After the End: Making U.S. Foreign Policy in the Post–Cold War World* (Durham, NC: Duke University Press, 1998).

58 See Steven W. Hook, "The White House, Congress, and the Paralysis of the State Department after the Cold War," in J. M. Scott, *After the End*, 305–29.

59 James M. Scott and A. Lane Crothers discuss the "shifting constellations" of power in U.S. foreign policymaking in "Out of the Cold: The Post–Cold War Context of U.S. Foreign Policy," in J. M. Scott, *After the End*, 1–25, especially p. 8.

60 Hook, "White House, Congress, and the Paralysis of the State Department," 318, 326. See also Warren Christopher, "American Diplomacy and the Global Environmental Challenges of the Twenty-First Century," speech at Stanford University, Apr. 9, 1996, available at Global Change Research Group, www.usgcrp.gov/usgcrp/documents/CWarren.html.

61 Hook, "White House, Congress, and the Paralysis of the State Department," 311–19.

62 U.S. Congress, 105 S. Res. 98, July 29, 1997 (Byrd-Hagel).

63 John H. Cushman Jr., "Intense Lobbying against Global Warming Treaty," *New York Times*, Dec. 7, 1997.

64 Ibid. On NAFTA and Kyoto, see David J. Hornsby, Alastair J. S. Summerlee, and Kenneth B. Woodside, "NAFTA's Shadow Hangs over Kyoto's Implementation," *Canadian Public Policy/Analyse de Politiques* 33:3 (Sept. 2007): 285–97. See also Renee G. Scherlen, "NAFTA and Beyond: The Politics of Trade in the Post–Cold War Period," in Scott, *After the End*, 258–385.

65 See Robert Repetto, "The Clean Development Mechanism: Institutional Breakthrough or Institutional Nightmare?" *Policy Sciences* 34:3/4 (2001): 303–27.

66 See Richard L. Sandor, Eric C. Bettelheim, and Ian R. Swingland, "An Overview of a Free-Market Approach to Climate Change and Conservation," *Philosophical Transactions: Mathematical, Physical, and Engineering Sciences* 360:1797 (Aug. 15, 2002): 1607–20; and Lawrence H. Goulder and Brian M. Nadreu, "International Approaches to Reducing Greenhouse Gas Emissions," in *Climate Change Policy: A Survey*, ed. Stephen H. Schneider, Armin Rosencranz, and John O. Niles, 115–50 (Washington, D.C.: Island Press, 2002).

67 Stavins, "What Can We Learn from the Grand Policy Experiment?"; Warwick J. McKibben and Peter J. Wilcoxen, "The Role of Economics in Climate Change Policy," *Journal of Economic Perspectives* 16:2 (Spring 2002): 107–29.

68 Scholars of environmental politics have yet to adequately address the origins and pejorative nature of the *command-and-control* term. Like *entitlements* in the debate over raising the debt ceiling in 2011, the *command-and-control* wording enabled economic conservatives to control the debate by controlling the language. See Stavins, "What Can We Learn from the Grand Policy Experiment?," 69; and Yergin, *Quest*, 485.

69 Bonnie G. Colby, "Cap and Trade Policy Challenges: A Tale of Three Markets," *Land Economics* 76:4 (Nov. 2000): 638–58.

EPILOGUE

1 Jonathan Zasloff and Jonathan M. Zasloff, "Massachusetts v. Environmental Protection Agency," *American Journal of International Law* 102:1 (Jan. 2008): 134–43.

2 U.S. Environmental Protection Agency, "EPA Denies Petition to Regulate Greenhouse Gas Emissions from Motor Vehicles," U.S. EPA Office of Public Affairs press release, Aug. 23, 2003, http://yosemite.epa.gov/opa/admpress .nsf/5dc6087040a927d3852572a000650c01/694c8f3b7c16ff6085256d900065fdad! OpenDocument.

3 Massachusetts v. Environmental Protection Agency, 415 F.3d 50, 367 U.S. App. D.C. 282 (D.C. Cir. 2005).

4 Massachusetts v. Environmental Protection Agency, 549 U.S. 497 (2007).

5 Ibid.

6 Felicity Berringer, "Group Seeks EPA Rules on Emissions from Vehicles," *New York Times*, Apr. 3, 2008. On the case, see Tyler Welti, "Massachusetts v.

EPA's Regulatory Interest Theory: A Victory for the Climate, Not Public Law Plaintiffs," *Virginia Law Review* 94:7 (Nov. 2008): 1751–85.

7   Joe Garofoli, "Gore Movie Reaching the Red States, Too," *San Francisco Chronicle*, July 8, 2006; "An Inconvenient Truth," *Box Office Mojo*, http://boxofficemojo.com/movies/?page=intl&id=inconvenienttruth.htm.

8   All quotations from Al Gore's movie are from Davis Guggenheim, dir., *An Inconvenient Truth* (Hollywood, CA: Paramount Pictures, 2006).

9   Naomi Oreskes, "Beyond the Ivory Tower: The Scientific Consensus on Climate Change," *Science* 306:5702 (Dec. 2004): 1686.

10 See, for example, A. Cunsolo Willox, S. Harper, J. Ford, K. Landman, K. Houle, and V. L. Edge, "From This Place and of This Place: Climate Change, Sense of Place, and Health in Nunatsiavut, Canada," *Social Science and Medicine* 75:3 (2012): 538–47; J. Ford, B. Smit, J. Wandel, M. Allurut, K. Shappa, H. Ittusujurat, and K. Qrunnut, "Climate Change in the Arctic: Current and Future Vulnerability in Two Inuit Communities in Canada," *Geographical Journal* 174:1 (2008): 45–62; and "Small Island States," in "Working Group II: Impacts, Adaptation, and Vulnerability," in Intergovernmental Panel on Climate Change, *Climate Change 2007: The Fourth Assessment Report of the Intergovernmental Panel on Climate Change* (Geneva: IPCC, 2007).

11 Hulme, *Why We Disagree about Climate Change* (Cambridge: Cambridge University Press, 2009), 334–37. Hulme borrows from Horst Rittel and Melvin Webber, "Dilemmas in a General Theory of Planning," *Policy Sciences* 4 (1973): 155–69.

12 Richard Lazarus has further distinguished climate change by emphasizing the issues of scale and discounting that characterize global warming, categorizing climate change not just as "wicked," but "super wicked." Richard Lazarus, "Super Wicked Problems and Climate Change: Restraining the Present to Liberate the Future," *Cornell Law Review* 94:5 (2009): 1153–1234; Kelly Levin, Steven Bernstein, Benjamin Cashore, and Graeme Auld, "Playing It Forward: Path Dependency, Progressive Incrementalism, and the 'Super Wicked' Problem of Global Climate Change," *IOP Conference Series: Earth and Environmental Science* 6 (2012), http://iopscience.iop.org/1755-1315/6/50/502002.

13 Stephen M. Wheeler, "State and Municipal Climate Change Plans: The First Generation," *Journal of the American Planning Association* 74:4 (Autumn 2008): 481–96.

14 Ibid., 482.

15 S. H. Schneider, S. Semenov, A. Patwardhan, I. Burton, C. H. D. Magadza, M. Oppenheimer, A. B. Pittock, et al., "Assessing Key Vulnerabilities and the Risk from Climate Change," in *Climate Change 2007: Impacts, Adaptation and Vulnerability; Contribution of Working Group II to the Fourth Assessment Report of the Intergovernmental Panel on Climate Change*, ed. M. L. Parry, O. F. Canziani, J. P. Palutikof, P. J. van der Linden, and C. E. Hanson (Cambridge: Cambridge University Press, 2007), 779–810.

16 See Stephen H. Schneider, "The Worst-Case Scenario," *Nature* 458 (Apr. 30, 2009): 1104–5; Stephen H. Schneider, "Abrupt Non-Linear Climate Change, Irreversibility and Surprise," *Global Environmental Change* 1:3 (Oct. 2004): 245–58; Schneider et al., *Climate Change 2007: Impacts, Adaptation and Vulnerability; Contribution of Working Group II*; and Joel B. Smith, Stephen H. Schneider, Michael Oppenheimer, Gary W. Yohe, William Hare, Michael D. Mastrandrea, Anand Patwardhan, et al., "Assessing Dangerous Climate Change through an Update of the Intergovernmental Panel on Climate Change (IPCC) 'Reasons for Concern,'" *Proceedings of the National Academy of Sciences USA* 106:11 (2009): 4133–37.

17 G. A. Meehl, T. F. Stocker, W. D. Collins, P. Friedlingstein, A. T. Gaye, J. M. Gregory, A. Kitoh, et al., "Global Climate Projections," in *Climate Change 2007: The Physical Science Basis; Contribution of Working Group I to the Fourth Assessment Report of the Intergovernmental Panel on Climate Change*, ed. S. Solomon, D. Qin, M. Manning, Z. Chen, M. Marquis, K. B. Averyt, M. Tignor, et al. (New York: Cambridge University Press, 2007).

18 UNFCCC, "Text of the Convention," 4 (Article III, Principle 1).

19 Wheeler assesses the depth and efficacy of these state and municipal commitments in "State and Municipal Climate Change Plans."

SELECTED BIBLIOGRAPHY

Ackerman, B. A., and R. B. Stewart. "Reforming Environmental Law: The Democratic Case for Market Incentives." *Columbia Journal of Environmental Law* 13:171 (1988): 153–69.

Agarwal, Anil, and Sunita Narain. *Global Warming in an Unequal World: A Case of Environmental Colonialism.* New Delhi: Centre for Science and Environment, 1990.

Agrawala, Shardul. "Context and Early Origins of the Intergovernmental Panel on Climate Change." *Climatic Change* 39:4 (Aug. 1998): 605–20.

Allen, Jonathan, ed. *March 4: Scientists, Students, and Society.* Cambridge, MA: MIT Press, 1970.

Alvarez, Luis W. *Alvarez: Adventures of a Physicist.* New York: Basic, 1987.

Alvarez, Luis W., Walter Alvarez, Frank Asaro, and Helen V. Michel. "Extraterrestrial Causes for the Cretaceous-Tertiary Extinction." *Science* 208:4448 (June 6, 1980): 1095–1108.

Arrhenius, Gustaf Olof Svante. "Oral History of Gustaf Olof Svante Arrhenius." Interview by Laura Harkewicz, Apr. 11, 2006. Scripps Institution of Oceanography Archives, La Jolla, CA. http://libraries.ucsd.edu/locations/sio/scripps-archives/resources/collections/oral.html.

Arrhenius, Svante. "On the Influence of Carbonic Acid in the Air upon the Temperature of the Ground." *London, Edinburgh, and Dublin Philosophical Magazine and Journal of Science* ser. 5 (Apr. 1896): 237–76.

———. *Worlds in the Making: The Evolution of the Universe.* Trans. H. Borns. New York: Harper & Brothers, 1908.

Badash, Lawrence. "Nuclear Winter: Scientists in the Political Arena," *Physics in Perspective* 3:1 (Mar. 2001): 76–105.

———. *A Nuclear Winter's Tale: Science and Politics in the 1980s.* Cambridge, MA: MIT Press, 2009.

———. *Science and the Development of Nuclear Weapons: From Fission to the Limited Test Ban Treaty, 1939–1963.* Atlantic Highlands, NJ: Humanities Press, 1995.

Barney, Gerald O., ed. *Global 2000 Report to the President: Entering the 21st Century.* Washington, D.C.: U.S. Government Printing Office, 1980.

Barrow, John C. "An Age of Limits: Jimmy Carter and the Quest for a National Energy Policy." In *The Carter Presidency: Policy Choices in the Post-New Deal Era*, ed. Gary M. Fink and Hugh Davis Graham, 158–223. Lawrence: University Press of Kansas, 1998.

Batchelor, G. K. *An Introduction to Fluid Dynamics*. Cambridge: Cambridge University Press, 1967.

Beck, Roy, and Leon Kolankiewicz. "The Environmental Movement's Retreat from Advocating U.S. Population Stabilization (1970–1998): A First Draft of History." *Journal of Policy History* 12:1 (2000): 123–56.

Berger, A. "Milankovitch Theory and Climate." *Review of Geophysics* 26:4 (1988): 624–57.

Bird, Kai, and Martin J. Sherwin. *American Prometheus: The Triumph and Tragedy of J. Robert Oppenheimer*. New York: Random House, 2005.

Bocking, Stephen. *Nature's Experts: Science, Politics, and the Environment*. New Brunswick, NJ: Rutgers University Press, 2006.

Bodansky, D. "The U.N. Framework Convention on Climate Change: A Commentary." *Yale Journal of International Law* 18:2 (1993): 451–558.

Boia, Lucien. *The Weather in the Imagination*. Trans. Roger Leverdier. London: Reaktion Books, 2005.

Bolin, Bert, and Erik Eriksson. "Changes in the Carbon Dioxide Content of the Atmosphere and Sea Due to Fossil Fuel Combustion" (1958). In *The Atmosphere and the Sea in Motion: Scientific Contributions to the Rossby Memorial Volume*, ed. Bert Bolin, 130–42. New York: Rockefeller Institute Press, 1959. http://wiki.nsdl.org/index.php/PALE:ClassicArticles/GlobalWarming/Article8.

Botkin, Daniel B. *Discordant Harmonies: A New Ecology for the Twenty-First Century*. Oxford: Oxford University Press, 1990.

Boulding, Kenneth. "The Economics of the Coming Spaceship Earth." In *Environmental Quality in a Growing Economy: Essays from the Sixth RFF Forum*, ed. Henry Jarrett, 3–14. Baltimore: Johns Hopkins University Press, 1966.

Bowen, Mark. *Censoring Science: Inside the Political Attack on Dr. James Hansen and the Truth of Global Warming*. New York: Plume, 2008.

Boyle, Robert H., and R. Alexander Boyle. *Acid Rain*. New York: Schocken Books, 1983.

Brammer, H. "Floods in Bangladesh: Geographical Background to the 1987 and 1988 Floods." *Geographical Journal* 156:1 (Mar. 1990): 12–22.

Brandt, Allan. *The Cigarette Century: The Rise, Fall, and Deadly Persistence of the Product that Defined America*. Boulder, CO: Basic Books, 2007.

Braudel, Fernand. *The Mediterranean and the Mediterranean World in the Age of Philip II*. Trans. Siân Reynolds. New York: Harper and Row, 1972.

Broecker, Wallace S. "Climatic Change: Are We on the Brink of a Pronounced Global Warming?" *Science* 189:4201 (Aug. 8, 1975): 460.

———. Interview by Spencer Weart, Lamont-Doherty Earth Observatory, NY,

Nov. 14, 1997. Niels Bohr Library and Archives, American Institute of Physics, College Park, MD. www.aip.org/history/ohilist/23909_1.html.

Broecker, Wallace S., David L. Thurber, John Goddard, Teh-Lung Ku, R. K. Matthews, and Kenneth J. Mesolella. "Milankovitch Hypothesis Supported by Precise Dating of Coral Reef and Deep-Sea Sediments." *Science* 159:3812 (Jan. 19, 1968): 297–300.

Brooks, C. E. P. *Climate through the Ages: A Study of the Climatic Factors and Their Variations.* 2nd rev. ed. New York: Dover Publications, 1970.

Brown, Lester. *By Bread Alone.* With Erik P. Eckholm. New York: Praeger, 1974.

Brown, Lester, Christopher Flavin, and Sandra Postel. *Saving the Planet: How to Shape an Environmentally Sustainable Global Economy.* New York: W. W. Norton, 1991.

Bryson, Reid E., and Thomas J. Murray. *Climates of Hunger: Mankind and the World's Changing Weather.* Madison: University of Wisconsin Press, 1977.

Bundy, McGeorge. *Danger and Survival: Choices about the Bomb in the First Fifty Years.* New York: Random House, 1988.

Burks, Alice Rowe. *Who Invented the Computer? The Legal Battle That Changed Computing History.* New York: Prometheus Books, 2003.

Calder, Nigel. *The Weather Machine.* New York: Viking, 1975.

Callendar, Guy Stewart. "The Artificial Production of Carbon Dioxide and Its Influence on Climate." *Quarterly Journal of the Royal Meteorological Society* 64 (1938): 223–40.

Carson, Rachel. *The Sea around Us.* New York: New American Library, 1954.

———. *Silent Spring.* Boston: Houghton Mifflin, 1962.

Cattell, J. McKeen. "The Organization of Scientific Men." *Scientific Monthly* 14:6 (June 1922): 568–78.

Chakrabarty, Dipesh. "The Climate of History: Four Theses." *Critical Inquiry* 35 (Winter 2009): 197–222.

Christiansen, Norman L., James K. Agee, Peter F. Brussard, Jay Hughes, Dennis H. Knight, G. Wayne Minshall, James M. Peek, et al. "Interpreting the Yellowstone Fires of 1988." *Bioscience* 39:10 (Nov. 1989): 678–85.

Christianson, Gale E. *Greenhouse: The 200-Year Story of Global Warming.* New York: Penguin Books, 1999.

Chung, Jan, and Jon Halliday. *Mao: The Unknown Story.* New York: Random House, 2005.

Coase, Ronald. "The Problem of Social Cost." *Journal of Law and Economics* 3 (Oct. 1960): 1–44.

Colby, Bonnie G. "Cap and Trade Policy Challenges: A Tale of Three Markets." *Land Economics* 76:4 (Nov. 2000): 638–58.

Cohen, Linda R., and Roger G. Noll. *The Technology Pork Barrel.* Washington, D.C.: Brookings Institution, 1991.

Commoner, Barry. *The Closing Circle: Nature, Man, and Technology.* New York: Alfred A. Knopf, 1971.

Conca, Ken, Michael Alberty, and Geoffrey D. Dabelko. *Green Planet Blues: Environmental Politics from Stockholm to Rio.* College Park, MD: Westview Press, 1995.

Conservation Foundation. *Implications of Rising Carbon Dioxide Content of the Atmosphere.* New York: Conservation Foundation, 1963.

Conway, Erik M. *Atmospheric Science at NASA: A History.* Baltimore: Johns Hopkins University Press, 2008.

———. *High Speed Dreams: NASA and the Technopolitics of Supersonic Transportation, 1945–1999.* Baltimore: Johns Hopkins University Press, 2005.

Cowling, Ellis B. "Acid Precipitation in Historical Perspective." *Environmental Science and Technology* 16:2 (Feb. 1982): 110A–123A.

Cronon, William, ed. *Uncommon Ground: Toward Reinventing Nature.* New York: W. W. Norton, 1995.

Crutzen, Paul J. "The Influence of Nitrogen Oxides on the Atmospheric Ozone Content." *Quarterly Journal of the Royal Meteorological Society* 96:408 (Dec. 1970): 320–25.

Crutzen, Paul J., and John W. Birks. "The Atmosphere after a Nuclear War: Twilight at Noon." *Ambio* 11:2/3 (Winter 1982): 114–25.

Cullather, Nick. "Miracles of Modernization: The Green Revolution and the Apotheosis of Technology." *Diplomatic History* 28:2 (Apr. 2004): 227–54.

Curtis, Carolyn, ed. *Before the Rainbow: What We Know about Acid Rain.* Washington, D.C.: Edison Electric Institute, 1980.

Daston, Loraine, and Peter Galison. *Objectivity.* New York: Zone Books, 2007.

Dawkins, Richard. *The Selfish Gene.* Oxford: Oxford University Press, 1976.

Davidson, Keay. *Carl Sagan: A Life.* New York: Wiley, 2000.

Day, Deborah. "Roger Randall Dougan Revelle Biography," 2008. Scripps Institution of Oceanography Archives, San Diego. www.scilib.ucsd.edu/sio/biogr/Revelle_Biogr.pdf.

Deacon, Robert T., and Paul Murphy. "The Structure of an Environmental Transaction: The Debt-for-Nature Swap." *Land Economics* 73:1 (Feb. 1997):1–24.

Devall, Bill, and George Sessions, eds. *Deep Ecology: Living as if Nature Mattered.* Salt Lake City: Peregrine Smith Books, 1985.

Development and Environment: Report and Working Papers of a Panel of Experts Convened by the Secretary-General of the United Nations Conference on the Human Environment (Founex, Switzerland, June 4–12, 1971). Paris: Mouton, 1972. Also known as the Founex Report.

Dotto, Lydia, and Harold Shriff. *The Ozone War.* Garden City, NY: Doubleday, 1978.

Druffel, Ann. *Firestorm: Dr. James E. McDonald's Fight for UFO Science.* With a foreword by Jacques Vallée. Columbus, NC: Wild Flower Press, 2003.

Dubos, Rene, and Barbara Ward. *Only One Earth: The Care and Maintenance of a Small Planet.* New York: W. W. Norton, 1972.

Dwiggings, Don. *The SST: Here It Comes Ready or Not.* New York: Doubleday, 1968.

Edwards, Paul. *A Vast Machine: Computer Models, Climate Data, and the Politics of Global Warming.* Cambridge, MA: MIT Press, 2010.

———. "The World in a Machine." In *Systems, Experts, and Computers: The Systems Approach in Management and Engineering, World War II and After,* ed. Agatha Hughes and Thomas Hughes, 221–53. Cambridge, MA: MIT Press, 2000.

Egan, Michael. *Barry Commoner and the Science of Survival.* Cambridge, MA: MIT Press, 2007.

Ehrlich, Paul R. *The Population Bomb.* New York: Ballantine Books, 1968.

Ehrlich, Paul R., and Anne H. Ehrlich. *Population, Resources, Environment: Issues in Human Ecology.* San Francisco: W. H. Freeman, 1970.

Ehrlich, Paul R., John Harte, Mark A. Harwell, Peter H. Raven, Carl Sagan, George M. Woodwell, Joseph Berry, et al. "Long-Term Biological Consequences of Nuclear War." *Science* 222:4630 (Dec. 23, 1983): 1293–1300.

Ehrlich, Paul R., and John P. Holdren. "Dispute." *Environment* 14 (Apr. 1972): 23–52.

Ehrlich, Paul R., Carl Sagan, Donald Kennedy, and Walter Orr Roberts. *The Cold and the Dark: The World after Nuclear War; Conference on the Long-Term Worldwide Consequences of Nuclear War, Washington, D.C., 1983.* New York: W. W. Norton, 1984.

Elichirigoity, Fernando. *Planet Management: Limits to Growth, Computer Simulation, and the Emergence of Global Spaces.* Evanston, IL: Northwestern University Press, 1999.

Fagan, Brian. *The Little Ice Age: How Climate Made History, 1300–1850.* New York: Basic Books, 2000.

——— *The Long Summer: How Climate Changed Civilization.* New York: Basic Books, 2004.

Feshbach, Murray. *Ecological Disaster: Cleaning Up the Hidden Legacy of the Soviet Regime.* New York: Twentieth Century Fund Press, 1995.

Finch, Boyd L. *Legacies of Camelot: Stewart and Lee Udall, American Culture, and the Arts.* Norman: University of Oklahoma Press, 2008.

Firor, John. Interview by Earl Droessler, June 26, 1990. UCAR/NCAR Oral History Project, UCAR/NCAR Archives, Boulder, CO.

Fitzgerald, Frances. *Way Out There in the Blue: Reagan, Star Wars, and the End of the Cold War.* New York: Simon and Schuster, 2000.

Flannery, Tim. *The Weather Makers: How Man Is Changing the Climate and What It Means for Life on Earth.* New York: Atlantic Monthly Press, 2005.

Fleming, James Roger. *The Callendar Effect: The Life and Times of Guy Stewart Callendar (1898–1964), the Scientist Who Established the Carbon Dioxide Theory of Climate Change.* Boston: American Meteorological Society, 2007.

———. "The Climate Engineers: Playing God to Save the Planet." *Wilson Quarterly* (Spring 2007): 46–60.

———. *Historical Perspectives on Climate Change*. Oxford: Oxford University Press, 1998.

Fleming, James Rodger, and Vladimir Jankovic, eds. "Klima." Special issue, *Osiris* 26:1 (2011).

Flippin, J. Brooks. *Conservative Conservationists: Russell Train and the Emergence of American Environmentalism*. Baton Rouge: Louisiana State University, 2006.

———. *Nixon and the Environment*. Albuquerque: University of New Mexico Press, 2000.

Forman, Paul. "Behind Quantum Electronics: National Security as a Basis for Physical Research in the United States." *Historical Studies of the Physical and Biological Sciences* 18:1 (1987): 149–229.

Forrester, Jay. *Industrial Dynamics*. Cambridge, MA: MIT Press, 1961.

———. *Urban Dynamics*. Cambridge, MA: MIT Press, 1969.

———. *World Dynamics*. Cambridge, MA: Wright-Allen Press, 1971.

Fourier, Joseph. "General Remarks on the Temperatures of the Globe and the Planetary Spaces." Trans. Ebenezer Burgess. *American Journal of Science* 32 (1837): 1–20.

Freymond, Jacques. "New Dimensions in International Relations." *Review of Politics* 37:4 (Oct. 1975): 464–78.

Friedberg, Aaron L. "Science, the Cold War, and the American State." *Diplomatic History* 20:1 (Winter 1996): 107–18.

Friedman, Thomas. *Hot, Flat, and Crowded: Why We Need a Green Revolution— and How it Can Renew America*. New York: Farrar, Straus, and Giroux, 2008.

Fromson, J. "A History of Federal Air Pollution Control." *Ohio State Law Journal* 30:3 (Summer 1969): 516–36.

Funtowicz, S. O., and J. R. Ravetz. "Three Types of Risk Assessment and the Emergence of Post-Normal Science." In *Social Theories of Risk*, ed. S. Krimsky and D. Golden, 251–73. Westport, CT: Greenwood, 1993.

Gaddis, John Lewis. *Strategies of Containment: A Critical Appraisal of Postwar American National Security*. Oxford: Oxford University Press, 1982.

Galison, Peter, and Bruce Hevly, eds. *Big Science: The Growth of Large-Scale Research*. Stanford, CA: Stanford University Press, 1992.

Gelbspan, Ross. *Boiling Point: How Politicians, Big Oil and Coal, Journalists, and Activists Have Fueled a Climate Crisis—and What We Can Do to Avert Disaster*. New York: Basic Books, 2004.

———. *The Heat Is On: The Climate Crisis, the Cover-up, the Prescription*. New York: Basic Books, 1998.

Glantz, Michael. *Saskatchewan Spring Wheat Production 1974: A Preliminary Assessment of a Reliable Long-Range Forecast*. Environment Canada Climatological Studies 33. Downsview, Ontario: Environment Canada, 1979.

———. "The Value of a Long Range Weather Forecast for the West African Sahel." *Bulletin of the American Meteorological Society* 58:2 (Feb. 1977): 150–58.

Glantz, Stanton A., John Slade, Lisa A. Bero, Peter Nanauer, and Deborah E. Barnes. *The Cigarette Papers*. Berkeley: University of California Press, 1996.

Gleick, James. *Chaos: Making a New Science*. New York: Viking, 1987.

Gore, Albert. *Earth in the Balance: Ecology and the Human Spirit*. Boston: Houghton Mifflin, 1992.

Gottlieb, Robert. *Forcing the Spring: The Transformation of the American Environmental Movement*. Washington, D.C.: Island Press, 1993.

Goulder, Lawrence H., and Brian M. Nadreu. "International Approaches to Reducing Greenhouse Gas Emissions." In *Climate Change Policy: A Survey*, ed. Stephen H. Schneider, Armin Rosencranz, and John O. Niles, 115–50. Washington, D.C.: Island Press, 2002.

Grattan-Guinness, I. *Joseph Fourier, 1768–1830: A Survey of His Life and Work*. Cambridge, MA: MIT Press, 1972.

Gray, C. Boyden, and David B. Rivkin Jr. "A 'No Regrets' Environmental Policy." *Foreign Policy* 83 (Summer 1991): 47–65.

Guggenheim, David, director. *An Inconvenient Truth*. Hollywood, CA: Paramount Pictures, 2006.

Guha, Ramachandra. *Environmentalism: A Global History*. New York: Longman Publishing Group, 1999.

———. *The Unquiet Woods: Ecological Change and Peasant Resistance in the Himalaya*. Berkeley: University of California Press, 1989.

Guston, David H. "Boundary Organizations in Environmental Policy and Science: An Introduction." *Science, Technology, and Human Values* 26:4 (Autumn 2001): 399–408.

Hagen, Joel. *An Entangled Bank: The Origins of Ecosystem Ecology*. New Brunswick, NJ: Rutgers University Press, 1992.

Hallgren, Elizabeth Lynn. *The University Corporation for Atmospheric Research and the National Center for Atmospheric Research, 1960–1970: An Institutional History*. Boulder, CO: National Center for Atmospheric Research, 1974. Also known as the Green Book.

Halpern, Shannon. *The United Nations Conference on Environment and Development: Process and Documentation*. Providence, RI: Academic Council for the United Nations System, 1992.

Hansen, James. Interview by Spencer Weart, Goddard Institute for Space Studies, New York City, Oct. 23, 2000. Niels Bohr Library and Archives, American Institute of Physics, College Park, MD. www.aip.org/history/ohilist/24309_1.html.

Hansen, James, D. Johnson, A. Lacis, S. Lebedeff, P. Lee, D. Rind, and G. Russell. "Climate Impact of Increasing Carbon Dioxide." *Science* 213:4511 (Aug. 28, 1981): 957–66.

Harper, Kristine. *Weather by the Numbers: The Genesis of Modern Meteorology*. Cambridge, MA: MIT Press, 2008.

Harrison, Mark. *Climates and Constitutions: Health, Race, Environment and British Imperialism in India, 1600–1850*. Oxford: Oxford University Press, 1999.

Hart, David M., and David G. Victor. "Scientific Elites and the Making of U.S. Policy for Climate Change Research, 1957–74." *Social Studies of Science* 23:4 (Nov. 1993): 643–80.

Hartman, Dennis L. *Global Physical Climatology*. San Diego: Academic Press, 1994.

Hayden, Carl Trumbull. *International Geophysical Year: Special Report Prepared by the National Academy of Sciences for the Committee on Appropriations of the United States Senate*. Washington, D.C.: U.S. Government Printing Office, 1956.

Hayes, Denis. *Rays of Hope: The Transition to a Post-Petroleum World*. New York: W. W. Norton, 1977.

Hays, Samuel P. *Beauty, Health, and Permanence: Environmental Politics in the United States, 1955–1985*. New York: Cambridge University Press, 1987.

———. *Conservation and the Gospel of Efficiency: The Progressive Conservation Movement, 1890–1920*. Cambridge, MA: Harvard University Press, 1959.

Head, Tom, ed. *Conversations with Carl Sagan*. Jackson: University Press of Mississippi, 2006.

Hecht, Alan D., and Dennis Tirpak. "Framework Agreement on Climate Change: A Scientific and Policy History." *Climatic Change* 29:4 (Apr. 1995): 371–402.

Herivel, John. *Joseph Fourier: The Man and the Physicist*. Oxford: Clarendon Press, 1975.

Herken, Gregg. *Brotherhood of the Bomb: The Tangled Lives and Loyalties of Robert Oppenheimer, Ernest Lawrence, and Edward Teller*. New York: Henry Holt, 2002.

———. "In the Service of the State: Science and the Cold War." *Diplomatic History* 24:1 (Winter 2000): 107–15.

Herring, George C. *America's Longest War: The United States and Vietnam, 1950–1975*. 2nd ed. New York: McGraw-Hill, 1986.

Hevly, Bruce. "Reflections on Big Science and Big History." In *Big Science: The Growth of Large-Scale Research*, ed. Peter Galison and Bruce Hevly, 355–63. Stanford, CA: Stanford University Press, 1992.

Hirst, Eric. "Review: *A New Prosperity: Building a Sustainable Energy Future* by Solar Energy Research Institute." *Policy Sciences* 14:2 (Apr. 1982): 198–202.

Hoggan, John. *Climate Cover-Up: The Crusade to Deny Global Warming*. With Richard Littlemore. New York: Greystone Books, 2009.

Holden, Barry. *Democracy and Global Warming*. London: Continuum, 2002.

Hornsby, David J., Alastair J. S. Summerlee, and Kenneth B. Woodside. "NAFTA's Shadow Hangs over Kyoto's Implementation." *Canadian Public Policy/ Analyse de Politiques* 33:3 (Sept. 2007): 285–97.

Horwitch, Mel. *Clipped Wings: The American SST Conflict*. Cambridge, MA: MIT Press, 1982.

Howe, Joshua P. "Getting Past the Greenhouse: John Tyndall and the Nineteenth-Century History of Climate Change." In *The Age of Scientific Naturalism: John Tyndall and His Contemporaries*, ed. Bernard Lightman and Michael Reidy, 33–49. London: Pickering and Chatto, 2014.

———. "The Stories We Tell." *Historical Studies of the Natural Sciences* 42:3 (2012): 244–54.

Hughes, Agatha, and Thomas Hughes, eds. *Systems, Experts, and Computers: The Systems Approach in Management and Engineering, World War II and After*. Cambridge, MA: MIT Press, 2000.

Hulme, Michael. *Why We Disagree about Climate Change: Understanding Controversy, Inaction, and Opportunity*. Cambridge: Cambridge University Press, 2009.

Imbrie, John, and Katherine Palmer Imbrie. *Ice Ages: Solving the Mystery*. Cambridge, MA: Harvard University Press, 1979.

Impact Team. *The Weather Conspiracy: The Coming of the New Ice Age*. New York: Ballantine Books, 1977.

Intergovernmental Panel on Climate Change. *Climate Change 2007: The Fourth Assessment Report of the Intergovernmental Panel on Climate Change*. Geneva: IPCC, 2007.

International Council of Scientific Unions (ICSU). *ICSU and Climate Science: 1962–2006 and Beyond*. Paris: ICSU, June 2006. www.icsu.org/publications/about-icsu/icsu-climate-science-2006.

International Council of Scientific Unions (ICSU), Scientific Committee on Critical Problems of the Environment (SCOPE), Commission on Monitoring. *Global Environmental Monitoring: A Report Submitted to the United Nations Conference on the Human Environment, Stockholm 1972*. Stockholm: ICSU-SCOPE, 1971.

Ivanova, Maria. "Looking Forward by Looking Back: Learning from UNEP's History." In *Global Environmental Governance: Perspectives on the Current Decade*, ed. Lydia Swart and Estelle Perry, 26–47. New York: Center for UN Reform Education, 2007.

Jaeger, Jill. *Developing Policies for Responding to Climate Change: A Summary of the Discussions and Recommendations of the Workshops Held in Villach (28 Sept.–2 Oct. 1987) and Bellagio (9–13 Nov. 1987) under the Auspices of the Beijer Institute, Stockholm*. Geneva: World Meteorological Organization; Nairobi: United Nations Environment Programme, 1988.

Jasanoff, Sheila. "Contested Boundaries in Policy-Relevant Science." *Social Studies of Science* 17:2 (May 1987): 195–230.

———. *The Fifth Branch: Science Advisors as Policymakers*. Cambridge, MA: Harvard University Press, 1990.

Johansen, Bruce E. *The Global Warming Desk Reference*. Westport, CT: Greenwood Press, 2002.

Johnston, Harold. "Reduction of Stratospheric Ozone by Nitrogen Oxide Catalysts from Supersonic Transport Exhaust." *Science* 173:3996 (Aug. 1971): 517–22.

Kaiser, David. *American Physicists and the Cold War Bubble*. Chicago: University of Chicago Press, forthcoming.

Kargon, Robert, Stuart W. Leslie, and Erica Schoenberger. "Far Beyond Big Science: Science Regions and the Organization of Research and Development." In *Big Science: The Growth of Large-Scale Research*, ed. Peter Galison and Bruce Hevly, 334–54. Stanford, CA: Stanford University Press, 1992.

Keeling, C. D. "The Suess Effect: ^{13}Carbon-^{14}Carbon Interrelations." *Environment International* 2 (1979): 229–300.

Kellogg, William W. Files. UCAR/NCAR Archives, Boulder, CO.

———. Interview by Ed Wolff and Nancy Gauss, Feb. 10, 1980. UCAR/NCAR Oral History Project, UCAR/NCAR Archives, Boulder, CO.

———. "Mankind's Impact on Climate: The Evolution of an Awareness." *Climatic Change* 10 (1987): 113–36.

Kellogg, William W. and Stephen H. Schneider. "Climate Stabilization: For Better or Worse." *Science* 186 (1974): 1163–72.

Kennan, George F. "To Prevent a World Wasteland: A Proposal." *Foreign Affairs* 48:3 (Apr. 1970): 401–13.

Kennedy, Paul. *The Parliament of Man: The Past, Present, and Future of the United Nations*. New York: Random House, 2006.

Kevles, Daniel J. "Cold War and Hot Physics: Science, Security and the American State." *Historical Studies of the Physical and Biological Sciences* 20:2 (1990): 239–64.

———. *The Physicists: The History of a Scientific Community in Modern America*. Cambridge, MA: Harvard University Press, 1995.

Kirk, Andrew. *Counterculture Green: The Whole Earth Catalog and American Environmentalism*. Lawrence: University Press of Kansas, 2007.

Kolbert, Elizabeth. *Field Notes from a Catastrophe: Man, Nature, and Climate Change*. New York: Bloomsbury, 2006.

Korsmo, Fae L. "The Genesis of the International Geophysical Year." *Physics Today* (July 2007): 38–43.

Kotkin, Stephen. *Armageddon Averted: The Soviet Collapse, 1970–2000*. Oxford: Oxford University Press, 2008.

Kukla, George J., and R. K. Matthews. "When Will the Present Interglacial End?" *Science* 178:4057 (Oct. 13, 1972): 190–91.

Ladurie, Emmanuel Le Roy. "History without People." In *The Territory of the Historian*, trans. Ben and Siân Reynolds, 285–341. Chicago: University of Chicago Press, 1979.

———. *Times of Feast, Times of Famine: A History of Climate Since the Year 1000*. Trans. Barbara Bray. New York: Noonday Press, 1971.

LaFeber, Walter. *America, Russia, and the Cold War, 1945–1996*. 8th ed. New York: McGraw-Hill, 1997.

Lahsen, Myanna. "Experiences of Modernity in the Greenhouse: A Cultural Analysis of a Physicist 'Trio' Supporting the Backlash against Global Warming." *Global Environmental Change* 18 (2008): 204–19.

Landsberg, Helmut E. "Climate as a Natural Resource." *Scientific Monthly* 63:4 (Oct. 1946): 293.

———. "Forum." *E.O.S.: Transactions of the American Geophysical Union* 58:3 (Mar. 1977): 122.

———. "Review: *The Genesis Strategy—Climate and Global Survival*." *E.O.S.: Transactions of the American Geophysical Union* 57:9 (Sept. 1976): 634–35.

Lanchbery, John, and David Victor. "The Role of Science in the Global Climate Negotiations." In *Green Globe Yearbook of International Co-operation and Development*, ed. Helge Ole Bergesen and Georg Parmann, 29–40. Oxford: Oxford University Press, 1995.

Lash, Jonathan. *A Season of Spoils: The Reagan Administration's Attack on the Environment*. New York: Pantheon Books, 1984.

Latour, Bruno. *The Politics of Nature: How to Bring the Sciences into Democracy*. Trans. Catherine Porter. Cambridge, MA: Harvard University Press, 2004.

———. *Science in Action*. Cambridge, MA: Harvard University Press, 1987.

Latour, Bruno, and Steve Woolgar. *Laboratory Life: The Construction of Scientific Facts*. Princeton, NJ: Princeton University Press, 1986.

Lauren, Paul Gordon. "Diplomats and Diplomacy of the U.N." In *The Diplomats, 1939–1979*, ed. Gordon A. Craig and Francis L. Lowenheim, 459–95. Princeton, NJ: Princeton University Press, 1994.

Lazarus, Richard. "Super Wicked Problems and Climate Change: Restraining the Present to Liberate the Future." *Cornell Law Review* 94:5 (2009): 1153–1234.

Leggett, Jeremy. *The Carbon War: Global Warming and the End of the Oil Era*. New York: Routledge, 2001.

Lieber, Keir A., and Daryl Press. "The Rise of U.S. Nuclear Primacy." *Foreign Affairs* 86:2 (Mar./Apr. 2006): 42–55.

Lilienfeld, Robert. *The Rise of Systems Theory: An Ideological Analysis*. New York: John Wiley and Sons, 1978.

Longino, Helen E. *Science as Social Knowledge: Values and Objectivity in Scientific Inquiry*. Princeton, NJ: Princeton University Press, 1989.

Lorenz, Edward. "Reflections on the Conception, Birth, and Childhood of Numerical Weather Prediction." *Annual Review of Earth and Planetary Sciences* 34 (May 2006): 37–45.

Lovins, Amory. *Soft Energy Paths: Toward a Durable Peace*. New York: Harper Collins, 1979.

Lundberg, B. K. O. "Supersonic Adventure." *Bulletin of the Atomic Scientists* 21 (Feb. 1965): 29–33.

MacCracken, Michael C., S. B. Idso, J. Hansen, D. Johnson, A. Lacis, S. Lebedeff, P. Lee, et al. "Climatic Effects of Atmospheric Carbon Dioxide [exchange of letters]." *Science* 220:4599 (May 20, 1983): 873–75.

Machta, Lester. Interview by Julius London, Oct. 31, 1993. AMS/UCAR Tape Recorded Interview Project, UCAR/NCAR Archives, Boulder, CO.

Maddox, John. *The Doomsday Syndrome*. New York: McGraw-Hill, 1972.

Maier, Charles S. *Dissolution: The Crisis of Communism and the End of East Germany*. Princeton, NJ: Princeton University Press, 1997.

Malone, Thomas F. Interview by Earl Droessler, Feb. 18, 1989. AMS/UCAR Tape Recorded Interview Project, UCAR/NCAR Archives, Boulder, CO.

Manabe, Sukiro, and Richard T. Weatherald. "Thermal Equilibrium of the Atmosphere with a Given Distribution of Relative Humidity." *Journal of the Atmospheric Sciences* 24:3 (May 1967): 241–59.

Manzione, Joseph. "Amusing and Amazing and Practical and Military: The Legacy of Scientific Internationalism in American Foreign Policy, 1945–1963." *Diplomatic History* 24:1 (Winter 2000): 21–55.

Marsh, George Perkins. *Man and Nature*. 1865. With a introduction by David Lowenthal and foreword by William Cronon. Seattle: University of Washington Press, 2003.

Martin, Brian. "Nuclear Winter: Science and Politics." *Science and Public Policy* 15:5 (Oct. 1988): 321–34

Martino, Joseph Paul. *Science Funding: Politics and Porkbarrel*. New Brunswick, NJ: Transaction Publishers, 1992.

Matthews, William H., William W. Kellogg, and G. D. Robinson, eds. *Inadvertent Climate Modification: Report of Conference, Study of Man's Impact on Climate (SMIC), Stockholm*. Boston: MIT Press, 1971.

Mauch, Christof, Nathan Stoltzfus, and Douglas R. Weiner, eds. *Shades of Green: Environmental Activism around the Globe*. Lanham, MD: Rowman and Littlefield Publishers, 2006.

McCartney, Scott. *ENIAC: The Triumphs and Tragedies of the World's First Computer*. New York: Walker, 1999.

McCloskey, Andre T. "The Stewardship of Planet Earth: Protecting Our Planet Twenty Years of Defending the Environment." *BIOS* 60:1/2 (Mar.–May 1989): 40–43.

McCormick, John. *Reclaiming Paradise: The Global Environmental Movement*. Bloomington: Indiana University Press, 1989.

McKibben, Bill. *The End of Nature*. New York: Random House, 1989.

McKibben, Warwick J., and Peter J. Wilcoxen. "The Role of Economics in Climate Change Policy." *Journal of Economic Perspectives* 16:2 (Spring 2002): 107–29.

McNeill, J. R. *Something New under the Sun: An Environmental History of the Twentieth-Century World*. New York: W. W. Norton, 2000.

Meadows, Donella H., Dennis L. Meadows, Jørgen Randers, and William W. Behrens III. *The Limits to Growth: A Report for the Club of Rome's Project on the Predicament of Mankind*. New York: Signet, 1972.

Meisler, Stanley. *United Nations: The First Fifty Years*. New York: Atlantic Monthly Press, 1995.

Meyer, David S. *A Winter of Discontent: The Nuclear Freeze Movement and American Politics*. New York: Praeger, 1990.

Michaels, Patrick. *Sound and Fury: The Science and Politics of Global Warming*. Washington, D.C.: Cato Institute, 1992.

Miller, Clark A., and Paul Edwards, eds. *Changing the Atmosphere: Expert Knowledge and Environmental Governance*. Cambridge, MA: MIT Press, 2001.

Molina, Mario J., and F. Sherwood Rowland. "Stratospheric Sink for Chlorofluoromethanes: Chlorine Atom-Catalysed Destruction of Ozone." *Nature* 249 (June 28, 1974): 810–12.

Morgan, Judith, and Neil Morgan. *Roger: A Biography of Roger Revelle*. San Diego: University of California, San Diego—Scripps Institution of Oceanography, 1996.

Naess, Arne. "Sustainable Development and Deep Ecology." In *Ethics of Environment and Development: Global Challenge, International Response*, ed. J. Ronald Engel and Joan Gibb Engel, 87–96. London: John Wiley, 1992.

———. "Sustainable Development and the Deep Ecology Movement." In *Sustainable Development: Theory, Policy, and Practice within the European Union*, ed. Susan Baker, Maria Kousis, Dick Richardson, and Steven Young, 61–71. London: Routledge, 1997.

Nash, Roderick Frazier. *Wilderness and the American Mind*. 4th ed. New Haven, CT: Yale University Press, 2001.

National Academy of Sciences, Carbon Dioxide Assessment Committee. *Changing Climate*. Washington, D.C.: National Academy of Sciences, 1983. Also known as the Nierenberg Report.

National Academy of Sciences, Climate Research Board. *Carbon Dioxide and Climate: A Scientific Assessment*. Washington, D.C.: National Academy of Sciences, 1979. Also known as the Charney Report.

National Academy of Sciences, Committee on Atmospheric Sciences. *The Atmospheric Sciences, 1961–1971*. Vol. 1, *Goals and Plans*. Washington, D.C.: National Academy of Sciences–National Research Council, 1962.

National Academy of Sciences, Committee on Atmospheric Sciences, Panel on Weather and Climate Modification. *Weather and Climate Modification: Problems and Prospects*. Washington, D.C.: National Academy of Sciences, 1966.

National Academy of Sciences, Geophysics Research Board. *Energy and Climate: Studies in Geophysics*. Washington, D.C.: National Academy of Sciences, 1977.

National Academy of Sciences, U.S. Committee for the Global Atmospheric Research Program, National Research Council. *Understanding Climatic Change: A Program for Action.* Washington, D.C.: National Academy of Sciences, 1975.

National Academy of Sciences–National Research Council, Panel on International Meteorological Cooperation. *Feasibility of a Global Observation and Analysis Experiment.* Washington, D.C.: National Academy of Sciences–National Research Council, 1966.

National Defense University. *Climate Change to the Year 2000: A Survey of Expert Opinion.* Washington, D.C.: National Defense University, 1978.

National Renewable Energy Laboratory. *25 Years of Research Excellence, 1977–2002.* Boulder, CO: NREL, 2002.www.nrel.gov/docs/gen/fy02/30845.pdf.

National Research Council. "Biological and Medical Effects of Nitrogen Oxide Emissions." In *Environmental Impact of Stratospheric Flight: Biological and Climatic Effects of Aircraft Emissions in the Stratosphere,* report by the Climatic Impact Committee for the National Academy of Sciences and the National Academy of Engineering, 35–50. Washington, D.C.: National Academy of Sciences, 1975.

National Research Council, CO₂/Climate Review Panel. *Carbon Dioxide and Climate: A Second Assessment.* Washington, D.C.: National Academy of Sciences, 1982.

National Research Council, Committee on Atmospheric Sciences. *Atmospheric Ozone Studies: An Outline for an International Observation Program; A Report by the Panel on Ozone to the Committee on Atmospheric Sciences National Academy of Sciences, National Research Council, Nov., 1965.* Washington, D.C.: National Academy of Sciences–National Research Council, 1966.

National Research Council, Committee on Climate and Weather Fluctuations and Agricultural Production, Board on Agricultural and Renewable Resources. *Climate and Food: Climatic Fluctuations and U.S. Agricultural Production.* Washington, D.C.: National Academy of Sciences, 1976.

National Research Council, Committee on Impacts of Stratospheric Change. *Halocarbons: Effects on Stratospheric Ozone.* Washington, D.C.: National Academy of Sciences, 1976.

National Research Council, Panel on Nitrates. *Nitrates: An Environmental Assessment.* Washington, D.C.: National Academy of Sciences, 1978.

National Research Council, Panel on Water and Climate. *Climate, Climatic Change, and Water Supply.* Washington, D.C.: National Academy of Sciences, 1977.

National Science Foundation. *Federal Funds for Research and Development Detailed Historical Tables: Fiscal Years 1956–1994.* NSF 94-331. Bethesda, MD: Quantum Research Corp., 1994.

———. *Preliminary Plans for a National Center for Atmospheric Research: Second Progress Report of the University Committee on Atmospheric Research.*

Washington, D.C.: National Science Foundation, Feb. 1959. www.ncar.ucar .edu/documents/bluebook1959.pdf. Also known as the Blue Book.

Neimark, Peninah, and Peter Rhoades Mott, eds. *The Environmental Debate: A Documentary History*. Westport, CT: Greenwood Press, 1999.

Nilsson, Sten, and David Pitt. *Protecting the Atmosphere: The Climate Change Convention and Its Context*. London: Earthscan Publications, 1994.

Novick, Peter. *That Noble Dream: The "Objectivity Question" and the American Historical Profession*. Cambridge: Cambridge University Press, 1988.

Oppenheimer, Michael. "Developing Policies for Responding to Climate Change." *Climatic Change* 15:1 (Apr. 1989): 1–4.

Oreskes, Naomi, and Erik M. Conway. "Challenging Knowledge: How Climate Change Became a Victim of the Cold War." In *Agnotology: The Making and Unmaking of Ignorance*, ed. Robert N. Proctor and Londa Schiebinger, 55–89. Stanford, CA: Stanford University Press, 2008.

———. *Merchants of Doubt: How a Handful of Scientists Obscured the Truth on Issues from Tobacco Smoke to Global Warming*. New York: Bloomsbury, 2010.

Oreskes, Naomi, Erik M. Conway, and Matthew Shindell. "From Chicken Little to Dr. Pangloss: William Nierenberg, Global Warming, and the Social Deconstruction of Scientific Knowledge." *Historical Studies in the Natural Sciences* 38:1 (Winter 2008): 109–52.

Oreskes, Naomi, and Ronald E. Doel. "The Physics and Chemistry of the Earth." In *Cambridge History of Science*, vol. 5, *Modern Physical and Mathematical Sciences*, ed. Mary Jo Nye, 538–60. Cambridge: Cambridge University Press, 2002.

Owen, Kenneth. *Concorde and the Americans: International Politics of the Supersonic Transport*. Washington, D.C.: Smithsonian Institution Press, 1997.

Palmer, Trevor. *Perilous Planet Earth: Catastrophes and Catastrophism through the Ages*. New York: Cambridge University Press, 2003.

Panjabi, Ranee K. L. "The South and the Earth Summit: The Development/ Environment Dichotomy." In *The Earth Summit at Rio: Politics, Economics, and the Environment*, 93–155. Boston: Northeastern University Press, 1997.

Parker, Sybil P. *McGraw-Hill Modern Scientists and Engineers*. New York: McGraw-Hill, 1980.

Parson, Edward A. *Protecting the Ozone Layer: Science and Strategy*. New York: Oxford University Press, 2003.

Paterson, Matthew. *Global Warming and Global Politics*. London: Routledge, 1996.

Patterson, James T. *Restless Giant: The United States from Watergate to Bush v. Gore*. New York: Oxford University Press, 2005.

Perkins, Sid. "Cooling Climate 'Consensus' of 1970s Never Was." *ScienceNews* 174:9 (Oct. 25, 2008): 5–6.

Peterson, Thomas C., William M. Connolley, and John Fleck. "The Myth of the 1970s Global Cooling Scientific Consensus." *Bulletin of the American Meteorological Society* 89:9 (Sept. 2008): 1325–37.

Plass, Gilbert N. "Carbon Dioxide and the Climate." *American Scientist* 44 (1956): 302–16.

———. "The Carbon Dioxide Theory of Climate Change." *Tellus* 8 (1956): 140–54.

Pomerance, Rafe. "The Dangers from Climate Warming: A Public Awakening." In *The Challenge of Global Warming*, ed. Dean Edwin Abrahamson, 259–69. Washington, D.C.: Island Press, 1989.

Popper, Karl. *Objective Knowledge: An Evolutionary Approach*. Oxford: Clarendon Press, 1972.

Porter, Theodore. *Trust in Numbers: The Pursuit of Objectivity in Science and Public Life*. Princeton, NJ: Princeton University Press, 1995.

Powell, James Lawrence. *The Inquisition of Climate Science*. New York: Columbia University Press, 2011.

———. *Dead Pool: Lake Powell, Global Warming, and the Future of Water in the West*. Berkeley: University of California Press, 2008.

Proctor, Robert. *Cancer Wars: How Politics Shaped What We Know and Don't Know about Cancer*. New York: Basic Books, 1995.

———. *Value-Free Science? Purity and Power in Modern Knowledge*. Cambridge, MA: Harvard University Press, 1991.

Pyne, Stephen. "Burning Questions and False Alarms about Wildfires at Yellowstone." *Forum for Applied Research and Public Policy* 4:2 (Summer 1989): 31–39.

Rabe, Barry G. *Statehouse and Greenhouse: The Emerging Politics of American Climate Change Policy*. Washington, D.C.: Brookings Institution Press, 2004.

Rabinowitch, Eugene. "International Cooperation of Scientists." *Bulletin of the Atomic Scientists* 2 (Sept. 1946): 1.

Rasool, S. I., and S. H. Schneider. "Atmospheric Carbon Dioxide and Aerosols: Effects of Large Scale Increases on Global Climate." *Science* 173:3992 (July 9, 1971): 138–41.

Repetto, Robert. "The Clean Development Mechanism: Institutional Breakthrough or Institutional Nightmare?" *Policy Sciences* 34:3/4 (2001): 303–27.

Revelle, Roger. Interview by Earl Droessler, Feb. 1989. American Institute of Physics, College Park, MD.

———. "Introduction: The Scientific History of Carbon Dioxide." In *The Carbon Cycle and Atmospheric CO₂: Natural Variations Archean to Present*, Geophysical Monograph 32, ed. E. T. Sundquist and W. S. Broecker, 1–4. Washington, D.C.: American Geophysical Union, 1985.

———. "Marine Bottom Samples Collected in the Pacific Ocean by the *Carnegie* on Its Seventh Cruise." Ph.D. dissertation, University of California, Berkeley, 1936.

———. "Preparation for a Scientific Career." Oral history by Sarah Sharp, 1984, Sio Reference Series No. 88, Nov. 1988. Regional Oral History Office, Bancroft Library, University of California, Berkeley.

Revelle, Roger, and Hans E. Suess. "Carbon Dioxide Exchange between Atmosphere and Ocean and the Question of an Increase of Atmospheric CO_2 during the Past Decades." *Tellus* 9 (1957): 18–27.

Rhodes, Richard. *The Making of the Atomic Bomb*. New York: Simon and Schuster, 1986.

Roan, Sharon L. *Ozone Crisis: The 15 Year Evolution of a Sudden Global Emergency*. New York: Wiley and Sons, 1989.

Roberts, Walter Orr. "Early History of the High Altitude Observatory," July 5, 1966. UCAR Oral History Project, UCAR/NCAR Archives, Boulder, CO.

———. Interview by David H. DeVorkin, July 26–28, 1983. Oral History Program, National Air and Space Museum of the Smithsonian Institution, Washington, D.C.

Robertson, Thomas. "'This Is the American Earth': American Empire, the Cold War, and American Environmentalism." *Diplomatic History* 32:24 (Sept. 2008): 561–84.

Robinson, George D. Interview by Earl Droessler, June 27–28, 1994. AMS/UCAR Tape Recorded Interview Project, UCAR/NCAR Archives, Boulder, CO.

Rochon, Thomas R., and David S. Meyer. *Coalitions and Political Movements: The Lessons of the Nuclear Freeze*. Boulder, CO: Lynne Rienner, 1997.

Rosenbaum, E. M. *Betrayal: The Untold Story of the Kurt Waldheim Investigation and Cover-Up*. With W. Hoffer. New York: St. Martin's Press, 1993.

Rosenbloom, Joshua. "The Politics of the American SST Programme: Origin, Opposition, and Termination." *Social Studies of Science* 11:4 (Nov. 1981): 403–23.

Rothman, Hal K. *The Greening of a Nation? Environmentalism in the United States since 1945*. Fort Worth, TX: Harcourt Brace, 1998.

Rothschild, Emma. "Forum: The Idea of Sustainability; Introduction." *Modern Intellectual History* 8:1 (Apr. 2011): 147–51.

———. "Maintaining (Environmental) Capital Intact." *Modern Intellectual History* 8:1 (Apr. 2011): 193–212.

Rozwadowski, Helen. *Fathoming the Ocean: The Discovery and Exploration of the Deep Sea*. Cambridge, MA: Harvard University Press, 2005.

———. *The Sea Knows No Boundaries: A Century of Marine Science under ICES*. Seattle: University of Washington Press, 2002.

Rusin, N., and L. Flit. *Man versus Climate*. Trans. Dorian Rottenberg. Moscow: Peace Publishers, 1960.

Sagan, Carl. "Nuclear War and Climate Catastrophe: Some Policy Implications." *Foreign Affairs* 62:2 (Winter 1983–84): 257–92.

Sale, Kirkpatrick. *The Green Revolution: The American Environmental Movement, 1962–1992*. New York: Hill and Wang, 1993.

Savage, James. *Funding Science in America: Congress, Universities, and the Politics of the Academic Pork Barrel*. New York: Cambridge University Press, 1999.

Schaller, Michael. *Reckoning with Reagan: America and Its President in the 1980s.* Oxford: Oxford University Press, 1980.

Schell, Jonathan. *The Fate of the Earth.* New York: Random House, 1982.

Schneider, Stephen H. Interview by Bob Chervin, Jan. 10–13, 2002. AMS/UCAR Tape Recorded Interview Project, UCAR/NCAR Archives, Boulder, CO.

———. *The Genesis Strategy: Climate and Global Survival.* With Lynn E. Mesirow. New York: Plenum Press, 1976.

———. *Global Warming: Are We Entering a Greenhouse Century?* San Francisco: Sierra Club Books, 1989.

———. *Laboratory Earth: The Planetary Gamble We Can't Afford to Lose.* New York: Basic Books, 1997.

———. "On the Carbon Dioxide-Climate Confusion." *Journal of the Atmospheric Sciences* 32:11 (Nov. 1975): 2060–66.

———. *The Primordial Bond: Exploring Connections Between Man and Nature through the Humanities and Sciences.* New York: Plenum Press, 1981.

———. *Science as a Contact Sport: Inside the Battle to Save Earth's Climate.* Washington, D.C.: National Geographic, 2009.

———. "The Worst-Case Scenario." *Nature* 458 (Apr. 30, 2009): 1104–5.

Schneider, Stephen H., William W. Kellogg, V. Ramanathan, Conway B. Leovy, and Sherwood B. Idso. "Carbon Dioxide and Climate [exchange of letters]." *Science* 210:4465 (Oct. 3, 1980): 6–8.

Schneider, Stephen H., and Randi Londer. *The Coevolution of Climate and Life.* San Francisco: Sierra Club Books, 1984.

Schneider, Stephen H., Armin Rosencranz, and John O. Niles, eds. *Climate Change Policy: A Survey.* Washington, D.C.: Island Press, 2002.

Scott, James C. *Seeing Like a State: How Certain Schemes to Improve the Human Condition Have Failed.* New Haven, CT: Yale University Press, 1998.

———. *Weapons of the Weak: Everyday Forms of Peasant Resistance.* New Haven, CT: Yale University Press, 1985.

Scott, James M., ed. *After the End: Making U.S. Foreign Policy in the Post–Cold War World.* Durham, NC: Duke University Press, 1998.

Sears, Paul B. *Deserts on the March.* 1937. 4th ed. Norman: University of Oklahoma Press, 1980.

Seitz, Russel. Letter. *Foreign Affairs* 63:4 (Spring 1984): 998–99.

Seidel, Stephen, and Dale Keyes. *Can We Delay a Greenhouse Warming?* 2nd ed. Washington, D.C.: Environmental Protection Agency, 1983.

Shabecoff, Philip. *A Fierce Green Fire: The American Environmental Movement.* Rev. ed. Washington, D.C.: Island Press, 2003.

Shackley, Simon, and Brian Wynne. "Representing Uncertainty in Global Climate Change Science and Policy." *Science, Technology, and Human Values* 21:3 (Summer 1996): 275–302.

Shiva, Vandana. "The Greening of the Global Reach." In *Global Ecology: A New*

Arena for Political Conflict ed. Wolfgang Sachs, 249–56. London: Zed Books, 1993.

———. *The Violence of Green Revolution: Third World Agriculture, Ecology, and Politics.* London: Atlantic Highlands, 1991.

Singer, S. Fred, ed. *Global Effects of Environmental Pollution: Symposium on Global Environmental Pollution.* New York: Springer-Verlag, 1970.

Singer, S. Fred, Cresson H. Kearny, R. P. Turco, O. B. Toon, T. P. Ackerman, J. B. Pollack, and C. Sagan. "On a 'Nuclear Winter.'" *Science* 227:4685 (Jan. 25, 1985): 356–444.

Smagorinsky, Joseph. "Global Atmospheric Modeling and the Numerical Simulation of Climate." In *Weather and Climate Modification,* ed. Wilmot Hess, 633–86. New York: Wiley, 1974.

———. Interview by John Young, May 16, 1986. AMS/UCAR Tape Recorded Interview Project, UCAR/NCAR Archives, Boulder, CO.

Smith, Gaddis. *Morality, Reason, and Power: American Diplomacy in the Carter Years.* New York: Hill and Wang, 1986.

Solar Energy Research Institute. *A New Prosperity: Building a Sustainable Future.* Andover, MA: Brickhouse Publishing, 1981.

Solow, Robert. "On the Intergenerational Allocation of Natural Resources." *Scandinavian Journal of Economics* 88 (Mar. 1986): 141–49.

Spangenburg, Ray, and Kit Moser. *Carl Sagan: A Biography.* Westport, CT: Greenwood Press, 2004.

Speth, James Gustave. *Red Sky at Morning: America and the Crisis of the Global Environment.* New Haven, CT: Yale University Press, 2004.

———, ed. *Worlds Apart: Globalization and the Environment.* Washington, D.C.: Island Press, 2003.

Speth, James Gustave, and Peter M. Haas. *Global Environmental Governance.* Washington, D.C.: Island Press, 2006.

Stanhill, Gerald. "Climate Change Science Is Now Big Science." *Eos, Transactions of the American Geophysical Union* 80:35 (1999): 396.

Stavins, Robert. "What Can We Learn from the Grand Policy Experiment? Lessons from SO_2 Allowance Trading." *Journal of Economic Perspectives* 12:3 (Summer 1998): 69–88.

Stevens, William K. *The Change in the Weather: People, Weather, and the Science of Climate.* New York: Dell Publishing, 1999.

Study of Critical Environmental Problems. *Man's Impact on the Global Environment: Assessment and Recommendation for Action; Report of the Study of Critical Environmental Problems (SCEP).* Cambridge, MA: MIT Press, 1970.

Suess, Hans E. "Radiocarbon Concentration in Modern Wood." *Science* 122:3166 (Aug. 2, 1955): 415–17.

Sullivan, Walter. *Assault on the Unknown: The International Geophysical Year.* New York: McGraw-Hill, 1961.

Swedin, Eric G., and David L. Ferro. *Computers: The Life Story of a Technology.* Baltimore: Johns Hopkins University Press, 2007.

Swedish Preparatory Committee for the UN Conference on the Human Environment (Bert Bolin, chairman). *Sweden's Case Study for the United Nations Conference on the Human Environment: Air Pollution across National Boundaries.* Stockholm: Norstadt and Sons, 1972.

Teller, Edward. "Widespread After-Effects of Nuclear War." *Nature* 310 (Aug. 16, 1984): 621–62.

Thomas, William Leroy, ed., *Man's Role in Changing the Face of the Earth.* Chicago: University of Chicago Press, 1956.

Thompson, Louis M. "Weather Variability, Climatic Change, and Grain Production." *Science* 188:4188 (May 9, 1975): 535–41.

Thompson, Philip, and Edward Lorenz. "Dialogue between Phil Thompson and Ed Lorenz," July 31, 1986. AMS/UCAR Tape Recorded Interview Project, UCAR/NCAR Archives, Boulder, CO.

Thompson, Starley L., and Stephen H. Schneider. "Nuclear Winter Reappraised." *Foreign Affairs* 64:5 (Summer 1986): 981–1005.

Trager, James. *Amber Waves of Grain: The Secret Russian Wheat Sales That Sent American Food Prices Soaring.* New York: Arthur Fields Books, 1973.

———. *The Great Grain Robbery.* New York: Ballantine Books, 1975.

Turco, Richard P. "Carl Sagan and Nuclear Winter." In *Carl Sagan's Universe,* ed. Yervant Terzian and Elizabeth Bilson, 239–46. New York: Cambridge University Press, 1997.

Turco, Richard P., O. B. Toon, T. Ackerman, J. B. Pollack, and C. Sagan. "Nuclear Winter: Global Consequences of Multiple Nuclear Explosions." *Science* 222:4630 (Dec. 23, 1983): 1283–92.

Turner, Fred. *From Counterculture to Cyberculture: Stewart Brand, the Whole Earth Network, and the Rise of Digital Utopianism.* Chicago: University of Chicago Press, 2008.

Tyndall, John. "Further Researches on the Absorption and Radiation of Heat by Gaseous Matter." 1862. In *Contributions to Molecular Physics in the Domain of Radiant Heat,* 69–121. New York: Appleton, 1873.

———. "On Radiation through the Earth's Atmosphere." *Philosophical Magazine* 4:125 (1862): 200–206.

———. "On the Absorption and Radiation of Heat by Gases and Vapours, and on the Physical Connection of Radiation, Absorption, and Conduction." *Philosophical Magazine* 4:22 (1861): 169–94, 273–85.

Ungar, Sheldon. "The Rise and (Relative) Decline of Global Warming as a Social Problem." *Sociological Quarterly* 33:4 (Dec. 1992): 483–501.

U.S. President's Science Advisory Committee, Environmental Pollution Panel. *Restoring the Quality of Our Environment: Report of the Pollution Panel, President's Science Advisory Committee.* Washington, D.C.: The White House, 1965.

van der Sluijs, Jeroen, Josee van Eijndhoven, Simon Shackley, and Brian Wynne. "Anchoring Devices in Science for Policy: The Case of Consensus around Climate Sensitivity." *Social Studies of Science* 28:2 (Apr. 1998): 291–323.

Victor, David G. *The Collapse of the Kyoto Protocol and the Struggle to Slow Global Warming.* Princeton, NJ: Princeton University Press, 2004.

Visser, Dana R., and Guillermo A. Mendoza. "Debt-for-Nature Swaps in Latin America." *Journal of Forestry* 92:6 (Nov. 1994): 13–16.

Wallace, Linda L., ed. *After the Fires: The Ecology of Change in Yellowstone National Park.* New Haven, CT: Yale University Press, 2004.

Waller, Douglas C. *Congress and the Nuclear Freeze: An Inside Look at the Politics of a Mass Movement.* Amherst: University of Massachusetts Press, 1987.

Wang, Zuoyue. *In Sputnik's Shadow: The President's Science Advisory Committee and Cold War America.* New Brunswick, NJ: Rutgers University Press, 2008.

Ward, Barbara. *Spaceship Earth.* New York: Columbia University Press, 1968.

Warde, Paul. "The Invention of Sustainability." *Modern Intellectual History* 8:1 (Apr. 2011): 153–70.

Washington, Warren M. *Odyssey in Climate Modeling, Global Warming, and Advising Five Presidents.* Morrisville, NC: Lulu.com, 2006.

Weart, Spencer. *The Discovery of Global Warming.* Cambridge, MA: Harvard University Press, 2003.

———. "From the Nuclear Frying Pan into the Global Fire." Bulletin of the Atomic Scientists 48:5 (June 1992): 18–27.

———. "Money for Keeling: Monitoring CO_2 Levels." *Historical Studies in the Physical and Biological Sciences* 37:2 (Mar. 2007): 435–52.

Weinberg, Alvin. "Impact of Large-Scale Science on the United States." *Science* 134:3473 (July 1961): 161–64.

———. *Reflections on Big Science.* Cambridge, MA: MIT Press, 1967.

Weinberger, Caspar W. *Fighting for Peace: Seven Critical Years in the Pentagon.* New York: Warner Books, 1990.

Weiner, Douglas. *A Little Corner of Freedom: Russian Nature Protection from Stalin to Gorbachev.* Berkeley: University of California Press, 1999.

Weiner, Jonathan. *The Next One Hundred Years: Shaping the Future of Our Living Earth.* New York: Bantam, 1990.

Welti, Tyler. "Massachusetts v. EPA's Regulatory Interest Theory: A Victory for the Climate, Not Public Law Plaintiffs." *Virginia Law Review* 94:7 (Nov. 2008): 1751–85.

Westad, Odd Arne. *The Global Cold War: Third World Interventions and the Making of Our Time.* New York: Cambridge University Press, 2005.

Wheeler, Stephen M. "State and Municipal Climate Change Plans: The First Generation." *Journal of the American Planning Association* 74:4 (Autumn 2008): 481–96.

Whitaker, John C. Papers, Staff Member Office Files, White House Central Files. Nixon Presidential Materials Project, College Park, MD.

————. *Striking a Balance: Environment and Natural Resources Policy in the Nixon-Ford Years*. Washington, D.C.: American Enterprise Institution for Public Policy Research, 1976.

Williams, Jerry L. *The Rise and Decline of Public Interest in Global Warming: Toward a Pragmatic Conception of Environmental Problems*. Huntington, NY: Nova Science Publishers, 2001.

Williams, William Appleman. *The Tragedy of American Diplomacy*. 50th anniversary ed. New York: W. W. Norton, 2009.

Wilson, Andrew. *The Concorde Fiasco*. Middlesex, U.K.: Penguin Books, 1973.

Wohl, C. G. "Scientist as Detective: Luis Alvarez and the Pyramid Burial Chambers, the JFK Assassination, and the End of the Dinosaurs." *American Journal of Physics* 75:11 (Nov. 2007): 968–77.

Woodwell, George M. "The Carbon Dioxide Question." *Scientific American* 238:1 (Jan. 1978): 34–43.

Woodwell, George M., R. H. Whittaker, W. A. Reiners, G. E. Likens, C. C. Delwiche, and D. B. Botkin. "The Biota and the World Carbon Budget." *Science* 199:4325 (Jan. 13, 1978): 141–46.

World Climate Program. *Report of the International Conference on the Assessment of the Role of Carbon Dioxide and of Other Greenhouse Gases on Climate Variations and Associated Impacts, Villach, Austria, 9–15 October, 1985*. WMO No. 661. Geneva: World Meteorological Organization, 1986.

World Commission on Environment and Development. *Our Common Future: Report on the World Commission on Environment and Development*. Oxford: Oxford University Press, 1987. Also known as the Brundtland Report.

World Meteorological Organization. *Proceedings of the World Climate Conference: A Conference of Experts on Climate and Mankind, 12–23 Feb., 1979*. WMO No. 537. Geneva: World Meteorological Organization, 1979.

————. *Proceedings of the World Conference on the Changing Atmosphere: Implications for Global Security, Toronto, Canada, June 27–30*. WMO No. 710. Geneva: World Meteorological Organization/United Nations Environment Programme, 1988.

————. *Report of the International Conference on the Assessment of the Role of Carbon Dioxide and of Other Greenhouse Gases in Climate Variations and Associated Impacts, Villach, Austria, 9–15 Oct., 1985*. WMO No. 661. Geneva: World Meteorological Organization, 1986.

Yamarella, Ernest Y., and Randal H. Ihara, eds. *The Acid Rain Debate: Scientific, Economic, and Political Dimensions*. Boulder, CO: Westview Press, 1985.

Yergin, Daniel. *The Quest: Energy, Security, and the Remaking of the Modern World*. New York: Penguin, 2011.

Zasloff, Jonathan, and Jonathan M. Zasloff. "Massachusetts v. Environmental Protection Agency." *American Journal of International Law* 102:1 (Jan. 2008): 134–43.

INDEX

financial resources, transfer of, 177–78
Firor, John, 52
"first-generation issues," 115
first-order effects, 56
fluorocarbons, 150, 152, 153, 186
food production: climate, the environment, and, 103–4. *See also* economic development and developing nations
"forcing function of knowledge," 9, 107–12, 147–48, 194
foreign aid, 179–82, 189
"forest principles," 176
Forrester, Jay, 71–72
Founex, meeting at, 81–82
Founex Report, 82, 86
Friedman, Thomas, 13, 100

G

Garwin, Richard, 50
general circulation modelers, 99
general circulation models (GCMs), 41, 102; Cold War and, 22, 25, 27; criticism of the theoretical nature of, 99; defined, 24; dynamic core of, 25; elements of, 25; and the flow of energy, 25; history and development of, 22, 24–25; Model II and, 130, 131; NCAR and, 32–33; NOAA and, 51; radiation and, 25, 33, 40. *See also* global circulation models
Genesis Strategy: Climate and Global Survival, The (Schneider and Mesirow), 101–2, 113
George C. Marshall Institute, 140
glacial meltwater. *See* sea-level rise
Glantz, Mickey, 115
"global," 82; limits to, 89–92
global circulation models, 32, 41. *See also* general circulation models
Global Climate Protection Act of 1987, 160

global cooling, 95–100, 104, 112, 141
global environment, 69; climate, systems science, and the, 69–76
Global Environment Fund, 181
Global Environmental Facility (GEF), 178, 181, 182
global environmental governance: framework established at Stockholm Conference, 69, 82, 83, 89–91; stages of development of, 151–52
Global Environmental Monitoring, 70–71, 74
Global 2000 Report to the President, 124, 125, 127, 146, 150–52
global warming: collective social and political failures on, 6–7; debate over, 6; defined, 14; historical perspective on, 95–99; hits the mainstream, 160–62; reasons we are behind the curve on, 204; terminology, 13–14; tragedy of the history of, 203. *See also specific topics*
"global weirding," 13, 100
globalism, a new kind of, 77–80, 89
Goddard Institute for Space Studies (GISS), 98, 100, 130, 131, 161
"good science": the battle over, 61–64; the continuing battle for, 100–10; standards, 113. *See also* science-first approach to advocacy
Gorbachev, Mikhail S., 175
Gore, Al, 127–29, 160, 171, 187, 203, 205; *An Inconvenient Truth,* 128, 201–3; on climate mitigation and economics, 183–84, 192, 195; congressional CO_2 hearings and, 127–31, 138, 143; DOE and, 128; influence of, 127–29; Nobel Peace Prize, 198, 203; testimony at Kyoto, 193
Gorsuch, Ann, 124
governance: bottom-up/multilevel, 206 (*see also* science-first approach to

advocacy). *See also* global environmental governance

government bureaucracies: scientists working within (vs. against), 115–17. *See also* federal bureaucracy

Grand Policy Experiment, 17–20, 44, 55, 192

Gravy, Wavy, 83

Gray, Boyden, 183

Green Revolution, 103

greenhouse gases (GHGs), 155, 156, 186; terminology, 165. *See also specific gases*

H

Hagel, Chuck, 189–90. *See also* Byrd-Hagel Resolution

halogenated hydrocarbons (halocarbons), 150, 245n11

hamartia, 9–10

Hansen, James, 98, 130–32, 161

Hare, F. Kenneth, 102

Harris, Robert, 124

Hayes, Denis, 60, 120–23, 235n12

Hayes, Philip, 109

Hays, Sam, 48, 124

heat waves, 160

Helms, Jesse, 189

Herter, Christian, 88–89

Hevly, Bruce, 31

Hickel, Walter, 83

Hog Farm Commune, 83

Hollman, Herb, 51–52

House Committee on Energy and Natural Resources, 127

House Committee on Science and Technology, 128

Hulme, Mike, 13–14, 204–5

I

Inconvenient Truth, An (film), 128, 201–3

Intergovernmental Negotiating Committee, 182

Intergovernmental Panel on Climate Change (IPCC), 12, 147–48, 155; Cold War and, 12; and the "forcing function of knowledge," 147–48; institutionalizing the primacy of science, 165–69; and institutionalizing the primacy of science, 165–69; intergovernmental character, 159; mission, 158; and the new politics of consensus, 162–65; Soviet Union and, 159, 163, 165, 168; UNEP, WMO, and, 158–60, 163, 166, 168; UNFCCC and, 158, 162, 163, 166, 169, 171, 180, 194, 204; working groups, 158–59, 163–68

International Council of Scientific Unions (ICSU), 21, 74, 155

International Geophysical Year (IGY), 20–22, 42

J

Jastrow, Robert, 140

Johnson, Lyndon B., 27, 49

Johnston, Harold, 54, 62

joint implementation, 187, 191

Joyce, J. Wallace, 20

K

Kasahara, Akira, 98

Keeling, Charles David, 3, 34, 39, 42, 67; 1963 article on CO_2, 34, 36–38, 55, 57, 82, 95, 145; annually oscillating CO_2 data collected by, 97; background and life history, 20, 37, 57; bosses and scientific superiors, 34; on climate change, 37; environmentalism and, 57; funding, 12, 20, 44; at Mauna Loa, 3, 5, 17, 20, 40; overview and characterizations of, 57; Roger

Ozbekhan, Hasan, 71
ozone (O_3), 149
ozone depletion, 84, 91, 115, 151, 152; causes, 150, 152, 153, 245n11 (*see also* supersonic transport); climate change and, 146, 148, 150–55, 157; George C. Marshall Institute and, 140; *Global 2000 Report to the President* and, 125, 146, 150–52; health consequences, 151, 152; independent scientific consensus and political action on, 153, 154; IPCC and, 148; Montreal Protocol and, 151–54, 160, 186–87; Mustafa Tolba and, 154, 155; nuclear winter and, 135–37; super-sonic transport (SST) and, 51–56, 150; Vienna Convention and, 154. *See also* World Conference on the Changing Atmosphere
ozone layer, 149–50. *See also* ozone depletion

P

Palme, Olof, 87
Panel of Experts on Development and Environment, 81–82
Pei, I. M., 28
Peterson, Russell, 125, 126
Pewitt, N. Douglas, 128, 235n19
Pollack, Jim, 136–37. *See also* TTAPS group
pollution, 38; CO_2 as pollutant, 199; defined, 38; paradigm of, 42
Pomerance, Rafe, 129
population growth, 105–6. *See also* *Limits to Growth*
precautionary principle, 184
PrepCom III (third UNCED prepara-tory meeting), 175
President's Science Advisory Commit-tee (PSAC), 37–38
Proctor, Robert, 61

R

radiation, 18
radiation budget models, 32, 33
radioactive gases, 36
radiocarbon dating, 18–19
Rasool, Ishtiaque, 97–99
Reagan, Ronald, 121, 122; administra-tion, 6, 121–24, 126, 128–34, 143, 144, 146, 154, 158–62; appointees, 124–26, 128; contempt for social and environmental science research, 119; disdain for environmental move-ment, 119; DOE and, 117, 119–21, 123, 124, 127, 128, 131; policies, 128, 129, 131, 134, 135, 137, 139, 140, 143, 144, 162, 202; transition from Carter to, 119–21, 124
"Reagan antienvironmental revolu-tion," 123–25; reaction to, 125–30
Reagan reaction, 125, 144–46
Reichelderfer, Francis, 24
Reilly, William, 180, 181, 185
resource management, efforts to incor-porate climate science into, 107–8
Revelle, Roger Randall Dougan, 16–22, 34, 38–42, 60, 97; 1957–58 Interna-tional Geophysical Year (IGY) and, 20–22; AAAS and, 110, 111; advisory group, 111; as advisory to National Research Council Committee on Meteorology, 30; on air pollution and CO_2 research, 36; atmospheric-monitoring program designed by, 20; background and career path, 35–36; on buffering mechanism, 19, 35; as chair of Committee on Climate, 110; as chair of National Academy of Sciences Commit-tee on the Biological Effects of Atomic Radiation, 35–36; as chair of PSAC's carbon dioxide group, 38; characterization of, 12, 36; Charles

David Keeling and, 4, 17, 20; on CO_2 exchange between oceans and atmosphere, 35; contributions, 17–18; framed CO_2 rise as natural experiment, 17, 19–20; funding and, 18, 20, 22, 28, 44, 112, 128; Hans Suess and, 18; lobbying, 112; overview, 16–17; predictions, 38, 97; Scripps Institution of Oceanography and, 12, 16, 18; studies, 116; *Tellus* article, 19, 35; testimony, 22, 34–35, 127–28

Richardson, Lewis Fry, 24

Rio Earth Summit. *See* United Nations Conference on Environment and Development

Rivkin, David, Jr., 183

Roberts, Walter Orr, 32, 52, 60

Rothschild, Emma, 172

Rowland, Sherwood, 150

S

Sagan, Carl, 243n116; Conference on the Long-Term Worldwide Biological Consequences of Nuclear War, 137–39; criticisms of, 140–43; "crude threshold" concept, 139–42; *The Genesis Strategy* and, 101; on nuclear winter and nuclear disarmament, 136–43; Paul Ehrlich and, 139, 140; publications, 137–40; Stephen Schneider and, 140–42. *See also* TTAPS group

Schaefer, Vincent, 26

Scharlin, Patricia, 105, 107

Schell, Jonathan, 135

Scheuer, James, 127–28

Schlesinger, James, 120

Schneider, Stephen H., 75, 100–103; on aerosols and climate change, 97–99; on assessment work vs. real work in climate science, 159; Carl Sagan and, 140–42; criticisms of, 101–2, 140–43;

DOE funding, 128; *The Genesis Strategy*, 101–2, 113; on impact of climate change, 207; on international efforts to study the global atmosphere, 75; NCAR and, 102–3, 138, 140–42; on nuclear winter and nuclear autumn, 140–42; policy recommendations, 102; television appearances, 101–3, 229n22; tension between political beliefs and professional responsibility, 145; testimony, 127–28; three-dimensional model, 141; TTAPS and, 140–42

science: big and broad, 27–32; climate change advocates' faith in, 10 (*see also* science-first approach to advocacy); politics and, 8–9; privileging and institutionalizing the primacy of, 165–69, 203–4

science-first approach to advocacy, 68, 134, 203–4; and AAAS of the 1970s, 110; limitations of, 7, 9, 10, 69, 203, 207; vs. mainstream environmental advocacy, 45–46; political change and, 9; political liabilities of, 95; and scientific advances, 7; scientists and, 9, 61; of SST, 45; top-down, 9, 114–17, 147, 194, 205, 207; of UNFCCC and Kyoto Protocol, 194, 206

science-to-policy models, 206

Scientific Committee on Problems of the Environment (SCOPE), 74–75

scientific consensus, 113, 132

scientific internationalism, 76

scientific models. *See* climate science models

scientist-as-advocate paradox, 8

scientists, working within (vs. against) executive bureaucracies, 115–17

Scott, James C., 6

Scoville, Anthony, 129

Scripps Institution of Oceanography, 12, 16, 18, 20

Thorsson, Inga, 70
Timeon, Peter, 165
Tolba, Mustafa, 153–55, 157–59
Toon, Brian, 136–37. *See also* TTAPS group
top-down approach. *See* science-first approach to advocacy
Train, Russell, 50, 84, 85, 88–89
Tsongas, Paul, 125–27
TTAPS group (Turco, Toon, Ackerman, Pollack, and Sagan), 136–42
Turco, Richard, 136–37. *See also* TTAPS group
Tyndall, John, 3–4, 209n3

U

Udall, Stewart, 48
uncertainty, scientific, 34, 95, 107, 157, 168, 185; about supersonic transport (SST), 53, 56; diminishing, 35; and economic precaution, 134, 187–88, 190, 193, 194; (over)emphasizing, xii, 42, 63, 96, 102, 142, 162, 164, 165, 183, 184, 187–88, 193, 194; "first-generation issues" and, 115; language and rhetoric of, xi, 66, 168, 190, 203; and regulation, 134, 199, 200
United Nations (U.N.), 178; "one world" approach to, 178–79
United Nations Conference on Environment and Development (UNCED/Rio Earth Summit): and the end of the Cold War, 174–80; goals/objectives, 166, 171, 175–79, 181; overview, 175; UNFCCC and, 166, 171, 175, 176, 181
United Nations (U.N.) Conference on the Human Environment (Stockholm Conference), 66, 68–70, 82–89, 176–77; AAAS and, 110; America's nongovernmental

organizations and, 104; anti-Western position, 88; atmospheric scientists and, 65, 70, 94; background and inspiration for, 70–71, 74; characterizations of, 68, 69; developing nations, economic development, and, 81–82, 86, 90, 94, 104, 174; Earth Summit compared with, 176, 185–86; Earthwatch and, 84, 91; endeavor to address human environment, 70; as failed attempt to include CO_2 in global governance, 92; follow-up conference (*see* United Nations Conference on Environment and Development); framework of global environmental governance established at, 69, 82, 83, 89–91; "global environment" and, 68; goals for, 77; Hog Farm Commune and, 83; and the limits to "global," 89–92; Maurice Strong and, 78, 79; new global vision and, 77, 79–80; nuclear disarmament and, 86, 135; "Only One Earth" slogan/theme, 56, 77; politicization of the global environment at, 89–90; preparatory documents for, 74, 76; reports on earth's large-scale environmental systems, 70–71; SCEP and, 75; scientific community at, 90–91; Sierra Club and, 105; SMIC and, 74; studies for, 71 (*see also specific studies*); Svante Oden and, 148; Sweden and, 70, 135, 148; U.N. and, 78; U.N. Environment Fund and, 85; Vietnam War and, 87
United Nations Economic and Social Council (ECOSOC), 70
United Nations (U.N.) Environment Fund, 85–86
United Nations Environment Programme (UNEP), 90, 91

WEYERHAEUSER ENVIRONMENTAL BOOKS

Loving Nature, Fearing the State: American Environmentalism and Antigovernment Politics before Reagan, by Brian Allen Drake

Whales and Nations: Environmental Diplomacy on the High Seas, by Kurkpatrick Dorsey

Pests in the City: Flies, Bedbugs, Cockroaches, and Rats, by Dawn Day Biehler

How to Read the American West: A Field Guide, by William Wyckoff

Behind the Curve: Science and the Politics of Global Warming, by Joshua P Howe.

WEYERHAEUSER ENVIRONMENTAL CLASSICS

The Great Columbia Plain: A Historical Geography, 1805–1910, by D. W. Meinig

Mountain Gloom and Mountain Glory: The Development of the Aesthetics of the Infinite, by Marjorie Hope Nicolson

Tutira: The Story of a New Zealand Sheep Station, by Herbert Guthrie-Smith

A Symbol of Wilderness: Echo Park and the American Conservation Movement, by Mark Harvey

Man and Nature: Or, Physical Geography as Modified by Human Action, by George Perkins Marsh; edited and annotated by David Lowenthal

Conservation in the Progressive Era: Classic Texts, edited by David Stradling

DDT, Silent Spring, and the Rise of Environmentalism: Classic Texts, edited by Thomas R. Dunlap

The Environmental Moment, 1968–1972, by David Stradling

The Wilderness Writings of Howard Zahniser, edited by Mark Harvey

CYCLE OF FIRE, BY STEPHEN J. PYNE

Fire: A Brief History

World Fire: The Culture of Fire on Earth

Vestal Fire: An Environmental History, Told through Fire, of Europe and Europe's Encounter with the World

Fire in America: A Cultural History of Wildland and Rural Fire

Burning Bush: A Fire History of Australia

The Ice: A Journey to Antarctica